T0134916

Shielding of Electromagnetic Waves

George M. Kunkel

Shielding of Electromagnetic Waves

Theory and Practice

 Springer

George M. Kunkel
Spira Manufacturing Corporation
San Fernando, CA, USA

ISBN 978-3-030-19240-2   ISBN 978-3-030-19238-9   (eBook)
https://doi.org/10.1007/978-3-030-19238-9

This Springer imprint is published by the registered company Springer Nature Switzerland AG
The registered company address is: Gewerbestrasse 11, 6330 Cham, Switzerland

Preface

In the late 1960s and early 1970s, I taught several classes in the UCLA Extension Department on Electromagnetic Interference Control. In one of my classes, I had a student with a Ph.D. degree in Physics from MIT. When I lectured on shielding, presenting the accepted theory, he informed me and the class that the theory did not comply to the basic laws of physics. During break time, I met with him for coffee to learn the details of his concerns.

The next several years, I spent a considerable amount of time performing a detailed analysis and testing in the effort to confirm or deny his concerns. What I found showed me undeniably that *he was right* and the foundational precepts in the traditional theory on electromagnetic shielding are inaccurate.

The major relevant findings are:

1. A 1.0 MHz wave (300 m long) cannot be inserted (let alone bounce back and forth) inside a 1.0-ohm barrier, 1.0 μm thick.
2. Using the theory (with the wave bounding back and forth inside the barrier), the energy leaving the barrier can exceed the energy entering the barrier by as much as 60 dB (one million times).
3. Tests have illustrated that there is not a reflection inside the barrier.

Since that time over 40 years ago, I have continued to intellectually scrutinize, test, and challenge these theories. At this writing, I have spent over 50 years as a design engineer, with the first 30 years spent making my living by accepting full responsibility for the design and testing of systems. There is not a single textbook written that describes what a radiated electromagnetic (EM) wave is and how the wave is generated. The purpose of this book is to supply the engineering community with clear definitions of what an electromagnetic wave is and how the wave is generated and the physics associated with the penetration of a wave into and through a shielding barrier and through an EMI gasketed seam. Additionally, understanding the physics associated with a wave penetrating a shielding barrier and a gasketed seam in the barrier is also crucial. I believe that the engineers of the world need to understand the penetration of a wave through shielding barriers and EMI gasketed seams in order to properly package the electronic systems of the future. The purpose of writing this book is to correct these problems.

San Fernando, CA   George M. Kunkel

Acknowledgments

Thanks to my industry colleagues and friends: Ron Brewer, Michael J. Oliver, and Douglas C. Smith for their assistance in reviewing and refining the ideas in this book. Thanks to my family Bonnie Paul, Michael Kunkel, Wendy Kunkel, and Lallie Kunkel for all their support in my work and life.

Contents

List of Figures

List of Tables

About the Author

George M. Kunkel has been an EMC design engineer for over 50 years of his life. He earned both his B.S. and M.S. degrees in engineering at UCLA. The first 30 years of his career was spent accepting full responsibility for the EMC design and test of systems. His professional responsibilities include being Chairman of the Technical Committee on Interference Control for an 18-year period (1969–1987) and of a Shielding Theory and Practice working group over a period of 6 years (2002–2008) of the EMC Society of the IEEE. He has written and presented over 100 technical articles at various IEEE-sponsored international symposia, edited professional publications, and also taught courses on EMC System Design at UCLA extension.

Chapter 1
Introduction

There are currently two accepted methods of estimating the attenuation of an electromagnetic wave through shielding barrier materials. This approximation is defined as Shielding Effectiveness. Both methods use wave theory and quasi-stationary assumptions. One uses Maxwell's equations to estimate the attenuation. The other method uses the "analogy" between wave theory (as applied to transmission lines) and the penetration of a wave through a barrier material.

Using Maxwell's equations can result in fairly accurate attenuation values along with the value of the E and H fields of a wave as it exits the barrier. However, the use of the equations is extremely difficult to use (requires as many as 13 equations) [1] where there are conditions and constraints associated with their use. One of the conditions is that results must comply with Stokes function (the sum total of all energy entering and leaving a given area must equal zero unless there is a sink or source of power) [2].

The method of choice by the electrical/electronic industry is the use of the "analogy" between transmission lines (as predicted by wave theory) and the penetration of a wave through a barrier material. This method does not comply with Stoke's function (or the basic laws of physics). The method consists of a series of equations[1] which are [3]:

Shielding effectiveness (SE) $= R + A + B$
Where R (Reflective loss) $= 20 \log (K + 1)^2/4K$,
$K = Z_{wave}/Z_{barrier}$
Z_W (wave impedance) is obtained through a set of equations associated with a wave generated by an electric or magnetic dipole antenna
Z_B (barrier impedance) $= (1 + j)/\sigma\delta$

$1 + j$ signifies that the inductance of the barrier material is equal to the resistance. For computational purposes: $(1 + j) = (2)^{1/2}$.

---

[1] Note: See Appendix C.

© Springer Nature Switzerland AG 2020
G. M. Kunkel, *Shielding of Electromagnetic Waves*,
https://doi.org/10.1007/978-3-030-19238-9_1

σ is the conductivity of the barrier material for a cubic meter of material in mhos/meter.[2]

δ is skin depth and represents the conductive area of an infinitely thick barrier. Skin effect attenuates a wave through a barrier material using the formula $e^{-t/\delta}$ (where t equals the thickness of a barrier in meters). Integrating $e^{-t/\delta}$ from zero (0) to infinity (∞) we obtain the following:

$$\int_0^\infty e^{-t/\delta}\, dt = \delta\left(1 - e^{-t/\delta}\right)$$

Setting $t = \infty$, $\delta\,(1 - e^{-t/\delta}) = \delta$.

Therefore, $(1 + j)/\sigma\delta$ is the impedance of an infinitely thick barrier.

The equation $(1 + j)/\sigma\delta(1 - e^{-t/\delta})$ will provide the barrier impedance for all barrier thicknesses.

A (absorption loss) = 20 log $e^{-t/\delta}$ where t = thickness of the barrier in meters.

"A" is defined as an absorption loss. Since there is not a power loss (an I^2R loss) it should be defined as an attenuation factor (it is actually a skin effect attenuation).

$$B\left(\text{Re} - \text{reflection coefficient}\right) = 20\log\left[\left(\frac{K-1}{K+1}\right)^2 1 - e^{-2t/\delta}\right]$$

In the literature [4], "B" is portrayed as a wave bouncing back and forth inside a shielding barrier material where the wave bouncing back and forth produces a gain in energy such that the power of the wave leaving the barrier can be greater than the power entering the barrier. It is actually a correction factor for assumptions made in the reflection loss equations when the assumptions are not valid.

There are two assumptions and one error associated with the Reflection Loss equation. The assumptions are (1) the barrier is infinitely thick (i.e., $Z_B = (1 + j)/\sigma\delta$ where δ represents the conductive area of an infinitely thick barrier and (2) the impedance of the barrier is less than the impedance of the wave (wave theory predicts that there is a loss when $Z_W < Z_B$, where in actual practice there is no loss). The error is that the theory predicts a reflection inside the barrier. In actual conditions, there is not a reflection.

The "20" log for the reflection loss equation implies that the power loss of the E and H fields of the incident wave is attenuated equally.

Wave theory as applied to a transmission line proposes that when Z_O (characteristic impedance of the transmission line) is greater than Z_L (load impedance) the voltage is reflected and when $Z_O < Z_L$ there is a reflected current. The analogy associated with transmission lines and the penetration of a wave through a barrier is:

---

[2] Note: "Mhos/meter" has been re-established as "Siemens". The term mhos/meter is retained for clarity purposes.

- When Z_B is less than Z_W (impedance of incident wave E_1/H_1) there is an E field reflection.
- And when Z_B is less than free space there is a H field reflection when the wave strikes the exit side.

Since wave theory predicts a power loss (or 10 log), 10 log "R" is to be applied to the E field reflection on the incident side of the barrier and 10 log is to be applied to the H field reflection on the exit side.

The H field reflection does not occur for the following reasons:

1. According to the literature [4], when the barrier is thin, the wave that enters the barrier bounces back and forth inside the barrier. If we have a 1.0 MHz wave enter a 1.0-ohm aluminum barrier, the wave, which is 300-m-long, is claimed to enter and bounce back and forth inside a barrier that is less than a micron thick. It is inconceivable that a 300-m-long wave can exist in a barrier a micron thick, and it's impossible that it has the ability to bounce back and forth inside the barrier.
2. Based on the theory, there is a reflection when a wave moves from one medium to another when the mediums are a different impedance as predicted by "Wave Theory." However, using the equation associated with R, A, and B there is not a reflection (no loss in shielding effectiveness) when $Z_W \leq Z_B$. If there is not a reflection under these conditions, it can be concluded that there will not be a reflection when the wave passes from the barrier into free space.
3. When the barrier is thick (greater than $2\pi\delta$), the wave does not reach the opposite (exit) wall so there cannot be an H field reflection.
4. Shielding effectiveness tests [5] have been performed on thin shield material (2.0 ohms) space blanket where an H field reflection loss was not detected.

Note: Shielding Effectiveness Theory $SE = R + A + B$ proposes that the incident wave that strikes a shielding barrier is a vector consisting of an E and H field. When the wave with $Z_B < Z_W$ there is a reflection and when the wave strikes the exit surface of the barrier there is another reflection due to a difference in the impedance of the mediums (Z_B vs. free space). This is not actually what occurs.

Rather, when a wave is impinged on a barrier, current (surface current density J_S) flows on the surface of the barrier. The current (J_S) in amps/meter generates an H field at right angles to the current, where the value of the H field in A/m is equal to that of the current density [6]. This H field creates a back EMF (voltage) in a direction opposite to that of the current density. This creates a force called "skin effect." The surface current density is attenuated as the current penetrates the barrier. The value of the current as a function of depth into the barrier is equal to $J_S e^{-d/\delta}$.

The value of the E and H fields on the incident surface of the barrier J_S are [6]:

$$E_S = J_S Z_B \left(V/m \right)$$

$$H_S = J_S(A/m)$$

The values of E and H at any depth "d" into the barrier are:

$$E_d = J_d Z_B = J_S e^{-d/\delta} Z_B$$

$$H_d = J_d$$

The value of the E and H fields when the depth "d" into the barrier is equal to the thickness of the barrier "t" are:

$$E_t = J_t Z_B = J_S e^{-t/\delta} Z_B$$

$$H_t = J_t$$

The power of the wave at depth t is equal to:

$$E_t H_t \cos 45° \left(\text{the current lags the voltage by } 45° \right)$$

Since there is no reflected loss at depth "t," the wave leaving the barrier is equal to E_t and H_t (which has been shown to be accurate through test) [5]. The power decreases as the square of the distance from the incident wave power source. The power of a wave at any distance from the barrier can be obtained. The power level can lead to the value of the E and H fields.

Summary

The use of Maxwell's equations to estimate the Shielding Effectiveness of a shield material by the average engineer is not possible because of a lack of training in their use. The use of the equations by engineers well versed in the use of the equations is difficult at best. As a result, the use of the equations (SE = R + A + B) provided to the engineering community is the method of choice. Unfortunately, this method is highly flawed, provides inaccurate results, and cannot be used to understand the physics associated with a wave penetrating a shielded barrier. The following is a short summary of a method that has proven to produce accurate results.

    When an electromagnetic wave is impinged on a shielded barrier, the current (displacement current) in the wave is coupled to the barrier in A/m. The current (surface current density J_s) which is coupled to the barrier is attenuated by skin effect as the current penetrates the barrier. The wave on the exit surface of the barrier (which is a function of the current) will radiate out from the exit surface. When there is a seam (or EMI gasketed seam) in the barrier, the current (J_s) will generate a voltage across the seam. The penetration of the wave at the seam is directly proportional to the voltage across the seam.

The assumption made with regard to the shielding theory described above is that the barrier is infinite in size. When the shielding barrier is a wall of an enclosure (box, cabinet, etc.), current standing waves will exist on the barrier. This standing wave can be as high as 24 dB when the source of the wave is on the outside of the barrier, and as high as 60 dB with an inside source. This means that the attenuation of the barrier material and the EMI gasketed seam will not be as high as predicted. However, by understanding the variables associated with the attenuation characteristics of a shield material or a gasket, a correction can be easily made.

References

1. Halme L, Annanpalo J (1992) Screening theory of metallic enclosures. IEEE, EMC International Symposium, Anaheim, CA
2. Sokolnikoff I, Redheffer R (1958) Mathematics and physics in modern engineering. McGraw Hill
3. Frederick Research Corp. (1962) Handbook of Radio Frequency Interference, vol 3, (Methods of Electromagnetic Interference Suppression). Frederick Research Corp., Wheaton, MD
4. Ott HW (1976) Noise reduction techniques in electronic systems, Wiley, New York
5. Broaddus Al, Kunkel G (1992) Shielding effectiveness testing of 1.0 Ohm aluminized Mylar, IEEE, EMC International Symposium, Anaheim, CA
6. Hallen E (1962) Electromagnetic theory. Wiley, New York

Chapter 2
Theory of Shielding

The theory of shielding [1] begins with the generation of an electromagnetic (EM) wave. This is accomplished with the use of two (2) of Maxwell's equations and class notes from a 1960 UCLA course on EM theory. The result is a wave consisting of lines of flux (defined as displacement current dD/dt) that radiates away from the source. This displacement current generates the E and H fields, where the value of the H field in A/m is equal to the value of the displacement current. When the wave is impinged onto a shielding barrier, the current (displacement current) in the wave is coupled to the barrier. The current coupled to the barrier (surface current density J_S) is attenuated by skin effect, where the values of the E and H field in the barrier are attenuated equally. The value of the E and H fields that exit the barrier is equal to that at the exit surface of the barrier. If there is an EMI gasketed seam in the shield, the surface current density J_S that crosses the seam will create a voltage across the seam. The value of the wave that penetrates the seam is directly proportional to that voltage [2].

References

1. Kunkel G (2016) A circuit theory approach to calculating the attenuation of shielding barriers, Interference Technology Magazine
2. IEEE EMCS Education Committee, Experiments Manual

© Springer Nature Switzerland AG 2020
G. M. Kunkel, *Shielding of Electromagnetic Waves*,
https://doi.org/10.1007/978-3-030-19238-9_2

Chapter 3
Generation of Electromagnetic Waves

Introduction

Electromagnetic waves are generated by two types of sources. These are high impedance sources and low impedance sources [1].

The low impedance sources consist of transformers, motors, and other devices that have a large number of ampere turns. These sources are typically low frequency and make up about 5% of the interference sources of concern. Electromagnetic interference generated by these sources is typically felt by the equipment in which they are housed or closely spaced equipment within a room or a cabinet.

The high impedance sources include radio, radar, GPS, WIFI, and other communication sources. These sources generate the EM environment that our electronic systems must be designed to operate in. Wires that contain high-frequency signals within electrical and electronic equipment are also a source of high impedance signals. These sources can add to the environment unless the EM waves radiating from the wires are contained within the system.

The frequency range for high impedance sources varies from 20 kHz to over 40 GHz. The power level from the radar can approach 5000 V per meter and is classified as high intensity radiated field (HIRF).

Generation of High Impedance Wave

Wires within an electrical or electronic equipment that carry high frequency signals generate lines of flux (defined as displacement current) between the wire and a ground plane. These lines of flux radiate out generating an H field at right angles to the lines of flux. The value of the H field in A/m is the same as the value of the electric flux density (or displacement current dD/dt) [2]. The value of the E field is a function of variables associated with the displacement current [3].

© Springer Nature Switzerland AG 2020
G. M. Kunkel, *Shielding of Electromagnetic Waves*,
https://doi.org/10.1007/978-3-030-19238-9_3

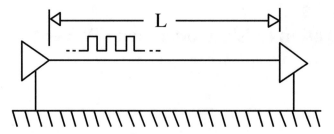

Fig. 3.1 Sending/receiving circuit above a ground plane. The wire (or PC card trace) acts as a transmitting antenna, radiating a high impedance EM wave

The generation of an electromagnetic wave can best be realized with the use of 2 of Maxwell's equations and class notes from William K. Hershberger's[1] 1960 UCLA, Engr. 117A Class on "Electromagnetic Theory." [2]

The two Maxwell's equations[2] are:

$$\text{rot}H = i + \mathrm{d}D/\mathrm{d}t \text{ and } E = D/\varepsilon_0\varepsilon$$

The equation "$\text{rot}H = i + \mathrm{d}D/\mathrm{d}t$" informs us that an H field is not only generated by current in a wire but also by displacement current in space. His other equation informs us that the E field is a function of electric flux density. When the flux density is generated by a sinusoidal current, D becomes $\mathrm{d}D/\mathrm{d}t$ or "displacement current."

W.K. Hershberger's lecture on the generation of an electromagnetic wave (using Maxwell's parallel plates in which Maxwell described the existence of displacement current) taught that the lines of flux (defined as displacement current) which radiate out from the edge of the parallel plates generate the E and H fields of a wave.

Figure 3.1 illustrates a sending/receiving circuit on a ground plane carrying a high frequency signal. The electromagnetic wave created by the circuit is a high impedance wave. The wave is identical (in terms of its impedance) to that generated by a transmission line pair, wire above a ground plane, electric dipole antenna, and the wave generated by a set of parallel plates described above.

---

[1] W.K. Hershberger's previous professional experience includes working for AT&T on the Transatlantic cable between the United States and England.

[2] See Appendix B.

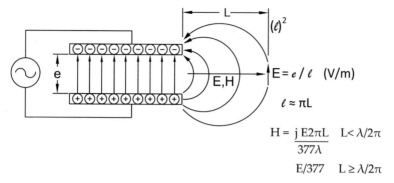

Fig. 3.2 Generation of an electromagnetic wave

Figure 3.2 illustrates a set of parallel plates, driven by a voltage source with the lines of flux (defined as displacement current) radiating out from the edge of the plates. These lines of flux represent the formation of a vertically polarized wave with a value in A/m. These lines of flux (displacement current) generate an H field (in A/m) at right angles to the displacement current where the value of the H field is equal to that of the displacement current [2]. The E field is a function of the displacement current. The relationship between the E and H fields[3] is [1]:

$$E/H = -j377\lambda / 2\pi L \quad L < \lambda / 2\pi$$
$$= 377 \quad L \geq \lambda / 2\pi \, (\text{m})$$

where λ = wave length of the frequency of interest.
= c/frequency (m).
c = speed of light = 3×10^8 m/s.
L = distance from source in meters.

References

1. Frederick Research Corp (1962) Handbook of radio frequency interference, vol 3 (Methods of Electromagnetic Interference Suppression). Frederick Research Corp., Wheaton, MD
2. Hershberger WK (1960) Class notes from UCLA class on electromagnetic theory
3. Hallen E (1962) Electromagnetic theory. Wiley, New York

---

[3] See Appendix C.

Chapter 4
Penetration of Electromagnetic Wave Through Shielding Barrier

Overview

The electromagnetic wave to be shielded by a barrier can be generated by low impedance and high impedances sources. When the impedance of a wave is less than the impedance of the barrier, shielding of the wave will not exist. When the impedance of the wave is 20 times (or more) then the impedance of the barrier, the wave is defined as a constant current source and the current coupled to the barrier is equal (in amperes/meter) to the value of the H field. When the impedance of the wave varies from that of the barrier to 20 times the impedance, the level of induced current can be predicted using a circuit theory analogy.

The circuit theory analogy is as illustrated in Fig. 4.1 and the following:
The value of the current J_S induced onto a barrier in terms of H_I is:

$$J_S = \left(\frac{Z_W}{Z_W + Z_B} \right) H_I$$

When $Z_W = Z_B$, the induced current = $\frac{1}{2} H_I$
$Z_W = 20 Z_B$, the induced current = 20/21 H_I

Penetration of Barrier [1]

When an electromagnetic (EM) wave is impinged onto a Shielding Barrier, the current (displacement current) in the wave is coupled to the barrier. Since the value of the H field is equal to the displacement current [2], the magnitude of the current in A/m coupled to the barrier is equal to that of the incident H field. The current (surface current density—J_S) that is induced into the barrier generates an H field at right

© Springer Nature Switzerland AG 2020
G. M. Kunkel, *Shielding of Electromagnetic Waves*,
https://doi.org/10.1007/978-3-030-19238-9_4

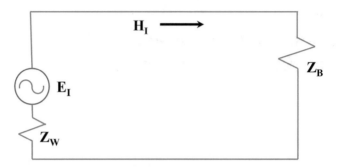

Fig. 4.1 Circuit theory analog to calculate current induced into barrier

angles to J_S. This H field creates a back EMF (voltage) opposite the direction of the current density. The voltage forces the current in the barrier to flow primarily on the surface of the barrier. This phenomenon is defined as "skin effect."

The value of the E and H fields on the surface of the barrier are [3]:

$$E_S = J_S Z_B \left(V / m \right)$$

$$H_S = J_S \left(A / m \right)$$

The current (due to skin effect) is attenuated as the current penetrates the barrier. The values of the E and H fields are attenuated equal to that of J_S. The values of the current, E and H fields at any depth "d" into the barrier are:

$$J_d = J_s e^{-d/\delta}$$

$$E_d = J_d Z_B$$

$$H_d = J_d$$

When d equals the thickness "t" of the barrier, the values of the E and H fields are:

$$J_t = J_s e^{-t/\delta}$$

$$E_t = J_t Z_B$$

$$H_t = J_t$$

The power of the wave at the surface "t" is equal to:

$$P_t = E_t H_t \cos 45° \left(\text{the current lags the voltage by} \, 45° \right)$$

The value of the wave as it leaves the barrier is equal to the wave on the exit surface of the barrier (see Fig. 4.2).

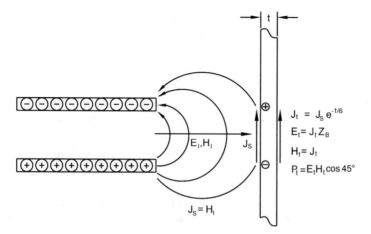

Fig. 4.2 Penetration of EM wave through shielding barrier

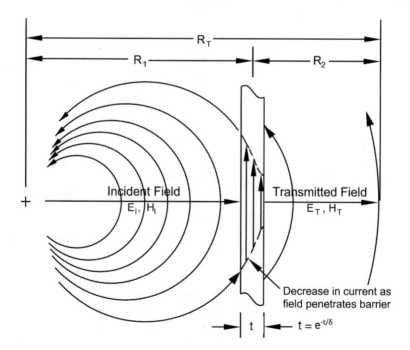

Fig. 4.3 Structural illustration of a wave penetrating a barrier

The power of the wave at any distance from a shielded barrier can be estimated using Fig. 4.3 and the following equation:

$$P_{RT} = P_{RI}\left(\frac{R_I}{R_T}\right)^2$$

Where the power is reduced as a square of the distance from the original source of power.

Figure 4.3 illustrates a wave striking a shielding barrier where: (1) the current (displacement current) in the wave is coupled to the barrier; (2) the current (surface current density "J_s") in the barrier is attenuated by skin effect; and (3) the wave (transmitted wave P_B) at the exit surface of the barrier radiates out of the barrier.

The contents of Tables 4.1 and 4.2 are used to obtain the values of "δ" and "Z_B."

Table 4.1 Equations and constants used to calculate the shielding of a barrier[a]

Skin depth δ

$$\delta = \left(\frac{2}{\sigma\mu\omega}\right)^{1/2} \text{ (meters)}$$

See Table 4.2 for the values of σ (conductivity) and μ (permeability)

$\omega = 2\pi$ frequency (Hz)

$$Z_B = \frac{(1+j)}{\sigma\delta\left(1-e^{-t/\delta}\right)} \text{ (ohms)}$$

For computational purposes $(1+j) = (2)^{1/2}$

t = thickness of barrier (meters)

The impedance of a wave from an electric and magnetic dipole [4]:

$E/H = -j377\lambda/2\pi L$
　　　for $L < \lambda/2\pi$ **High** impedance (**electric**) dipole antenna

$E/H = j377(2\pi L)/\lambda$
　　　for $L < \lambda/2\pi$ **Low** impedance (**magnetic**) dipole antenna

$E/H = 377$ for $L \geq \lambda/2\pi$ both antennas

L = distance from antenna (meters)

$\lambda = c/$frequency

c = speed of light = 3×10^8 (meters/second)

[a]See Appendix C

Table 4.2 Relative conductivity and permeability of metals [5–6]

	Metal	Relative conductivity (σ_r)	Relative permeability (μ_r)
1	Silver	1.05	1
2	Copper	1.00	1
3	Gold	0.70	1
4	Aluminum	0.61	1
5	Magnesium	0.38	1
6	Brass	0.35	1
7	Tungsten	0.31	1
8	Cadmium	0.23	1
9	Nickel	0.23	100
10	Iron	0.17	1000
11	Steel	0.17	1000
12	Tin	0.15	1
13	Lead	0.08	1
14	Monel	0.04	1
15	Manganese	0.04	1
16	Titanium	0.04	1
17	Stainless steel (430)	0.02	500

Notes: $\sigma = \sigma_c\,\sigma_r$ $\sigma_c = 5.82 \times 10^7$ (mhos/m) $\mu = \mu_o\,\mu_r$ $\mu_o = 4\pi \times 10^{-7}$ (H/m)
mhos/m has been redefined as "siemens" but is used here for clarity

Example: The electronics inside a drone made of fiberglass is protected from EMI by a plating of copper 0.001 in. thick. The attenuation of a plane wave having a field strength of 377 V/m ($H = 1.0$ A/m) at 2 GHz can be obtained as follows:

Step 1: Obtain the values of σ and μ from Table 4.2

$$\sigma = \sigma_{copper}\,\sigma_r = 5.82 \times 10^7 \left(\text{mhos} / \text{m}\right)$$

$$\mu = \mu_o\,\mu_t = 4\pi \times 10^{-7} \left(\text{H} / \text{m}\right)$$

Step 2: Calculate δ

$$\delta = \left(\frac{2}{\sigma\mu\omega}\right)^{\!\!1/2} = 2.0862 \times 10^{-6}$$

Step 3: Calculate Z_B

$$t\left(\text{meters}\right) = \left(0.001\,\text{in.}\right) / \left(29.37\,\text{in.} / \text{m}\right) = 2.54 \times 10^{-5} \left(\text{meters}\right)$$

$$Z_B = \frac{(1+j)}{\sigma\delta\left(1 - e^{-t/\delta}\right)} = 0.0116\,\text{Ohms}$$

Step 4: Document Surface Current Density J_s

$$J_s = H_I \left(H \text{ field of wave incident on barrier} \right) = 1.0 \, \text{A/m}$$

Step 5: Calculate Loss due to Skin Effect (LSE)

$$\text{LSE} = e^{-t/\delta} = 5.1566 \times 10^{-6}$$

Step 6: Obtain values of E and H at exit surface of barrier

$$E_t = J_s Z_B e^{-t/\delta} = 6.0060 \times 10^{-8} \, \text{V/m}$$

$$H_t = J_s e^{-t/\delta} = 5.1564 \times 10^{-6} \, \text{A/m}$$

$$P_t = E_t H_t \cos 45° = 2.1902 \times 10^{-13}$$

Step 7: Calculate the Shielding Attenuation (SA) for the E and H fields and power in dB

$$E_A \text{ field} (\text{dB}) = 20 \log E_I / E_t - 20 \log \left(377 / 6.0060 \times 10^{-8} \right) = 196 \text{dB}$$

$$H_A \text{ field} (\text{dB}) = 20 \log H_I / H_t - 20 \log \left(1.0 / 5.1564 \times 10^{-6} \right) = 106 \text{dB}$$

$$P_A \text{ field} (\text{dB}) = 10 \log P_I / P_t - 10 \log \left(377 / 2.1902 \times 10^{-13} \right) = 152 \text{dB}$$

References

1. Kunkel G (2016) A circuit theory approach to calculating the attenuation of shielding barriers, Interference Technology Magazine
2. Hershberger WK (1960) Class notes from UCLA class on electromagnetic theory
3. Hallen E (1962) Electromagnetic theory. Wiley, New York
4. Frederick Research Corp (1962) Handbook of radio frequency interference, vol 3 (Methods of Electromagnetic Interference Suppression). Frederick Research Corp., Wheaton, MD
5. Ott HW (1976) Noise reduction techniques in electronic systems. Wiley, New York
6. White DRJ (1973) Shielding, Don White Consultants Inc., Germantown, MD

Chapter 5
Penetration of Electromagnetic Wave Through EMI Gasketed Seam

Theory of Penetration [1]

Electromagnetic leakage through seams (and EMI gasketed seams) in shielded enclosures occurs primarily as a result of the currents which cross the seams.

Such crossing causes a voltage to appear on the far side of the seam. EM leakage via the seam is directly proportional to this (transfer) voltage.

In shielding theory, the seam is characterized in terms of its transfer impedance as follows:

$$Z_T = V / J_S$$

Z_T = Transfer impedance of seam (Ω meters)

V = Transfer Voltage across seam (or EMI gasketed seam)

J_S (surface current density) equals density or current which crosses the seam (A/m)

Penetration of EMI Gasketed Seam [2]

Figure 5.1 illustrates an EM wave impinged onto an EMI gasketed seam in a shielded enclosure.

The contents of Fig. 5.1 illustrate the following:

1. $J_S = H_I$ is the value of the current induced onto the barrier (A/m).
2. e is the voltage across the gaskets = $J_s Z_T$ (volts).
3. The value of the E field at a distance "L" from the gasketed seam.
4. The value of the H field at a distance of L from the gasketed seam.[1]

---

[1] See Appendix C

© Springer Nature Switzerland AG 2020
G. M. Kunkel, *Shielding of Electromagnetic Waves*,
https://doi.org/10.1007/978-3-030-19238-9_5

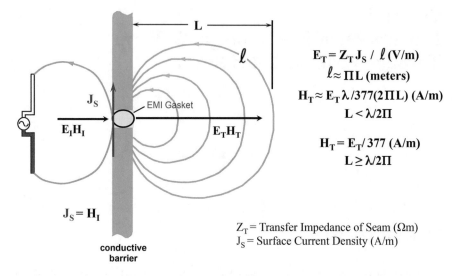

$$E_T = Z_T J_S / \ell \ (\text{V/m})$$
$$\ell \approx \Pi L \ (\text{meters})$$
$$H_T \approx E_T \lambda / 377 (2\Pi L) \ (\text{A/m})$$
$$L < \lambda/2\Pi$$

$$H_T = E_T / 377 \ (\text{A/m})$$
$$L \geq \lambda/2\Pi$$

$$J_S = H_I$$

Z_T = Transfer Impedance of Seam (Ωm)
J_S = Surface Current Density (A/m)

Fig. 5.1 Wave impinged on gasketed seam

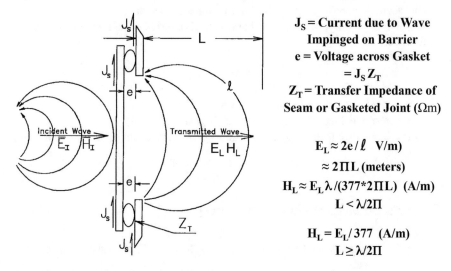

J_S = **Current due to Wave Impinged on Barrier**
e = Voltage across Gasket
$$= J_S Z_T$$
Z_T = **Transfer Impedance of Seam or Gasketed Joint (Ωm)**

$$E_L \approx 2e/\ell \ \text{V/m}$$
$$\approx 2\Pi L \ (\text{meters})$$
$$H_L \approx E_L \lambda/(377 \ast 2\Pi L) \ (\text{A/m})$$
$$L < \lambda/2\Pi$$

$$H_L = E_L / 377 \ (\text{A/m})$$
$$L \geq \lambda/2\Pi$$

Fig. 5.2 EMI gasketed maintenance cover

Figure 5.2 illustrates an EM wave impinged on an EMI gasketed cover.
What is illustrated in Fig. 5.2 is:

1. J_S crosses the gasketed seam at both the top and bottom of the cover generating 2 sources of voltage "e."
2. An estimate of the value of the E field at a distance L from the cover.

3. The value of the H field at distance L as a function of the E field.

Note: The wave leaving a seam (or EMI gasketed seam) is a low impedance wave. The value of the H field is based on the assumption that the impedance of the wave leaving the gasketed seam is identical to a wave generated by a magnetic dipole antenna.

Example: EMI Gasketed Cover

An electronic system inside the wheel well of an airplane with an EMI gasketed cover is exposed to a radar beam of 3770 Volts/meter at 2 GHz upon landing and taking off from an airfield. The EMI gasketed joint has a Z_T of 10^{-4} ohm-meters. The value of the E and H fields at a printed circuit card 0.05 m from the cover can be obtained using Fig. 5.3 and the following:

The field strength of 3770 V/m has an H field of 10.0 A/m. The surface current density J_S coupled to the gasketed joint is equal to the value of the H field and is 10.0 A/m.

The value of the gasketed joint is 10^{-4} ohm-m.

The voltage "e" across the gasketed joint is equal to $J_S Z_T = 10^{-3}$ (volts).

The value of the E field 0.05 m from the gasketed joint equals c/ℓ.

where $\ell = 0.05\,\pi = 0.1571$ m

$E = 10^{-3}/0.1571 = 6.366 \times 10^{-3}$ V/m

$H = E/377 = 1.69 \times 10^{-5}$ A/m

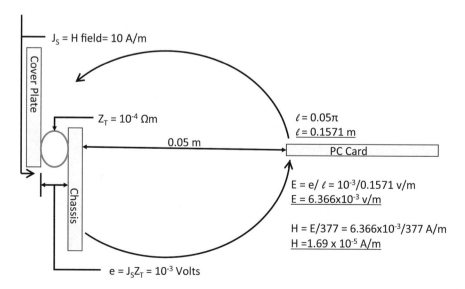

Fig. 5.3 EMI gasketed cover

References

1. IEEE EMCS Education Committee, Experiments Manual
2. Kunkel G (2018) Testing of EMI gaskets. Evaluation Engineering, June 2018

Chapter 6
Radiated Field Strength from Radiating Elements

The radiating elements are specifically configured wires attached to a thin fiberglass board measuring 20 in. high by 30 in. wide. The purpose of the test is to verify (1) that the measured H field is generated by current in the wave defined as "displacement current" and (2) that the wave impedance (E/H) from high impedance sources is equal.

The configured radiating elements are:

1. Loaded electric dipole antenna having the following characteristics: (a) a BNC connector located in the center of the 20 by 30 in. board; (b) two nine (9) in. long wires are attached to the BNC connector with one wire situated in the plus "y" direction and the other in the minus "y" direction; and (c) 29 in. long wires are attached to the ends of the wire leading to the BNC connector (see Fig. 6.1).
2. Two 29-in. long wires are attached to a fiberglass board 1 mm apart (see Fig. 6.2). One end of the wires is attached to a BNC connector and the other to a 50-ohm resistor.
3. Two wires 29 in. long separated 0.20 in. apart where the result is a transmission line pair having a characteristic impedance of 360 Ohms. One end is attached to a BNC connector and the other end to a 360-ohm resistor (see Fig. 6.3).

Power over the frequency range of 100 kHz to 1.0 MHz is supplied to the BNC connector of each of the radiating elements. The value of the E and H fields is recorded for each of the radiating elements. Figures 6.4, 6.5, and 6.6 illustrate the recorded data.

Table 6.1 tabulates the measured results of the E and H fields from the radiating elements. As can be observed, the impedance of the wave (within the accuracy limits of the test setup) is similar illustrating that the impedance from various high impedance sources is equal.

© Springer Nature Switzerland AG 2020
G. M. Kunkel, *Shielding of Electromagnetic Waves*,
https://doi.org/10.1007/978-3-030-19238-9_6

Fig. 6.1 Electric dipole
antenna

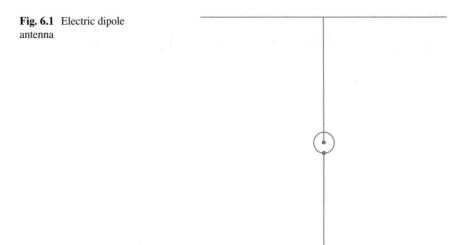

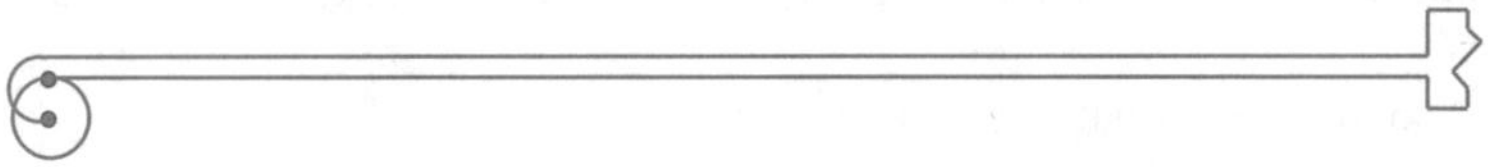

Fig. 6.2 Closely spaced wire pair

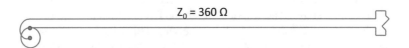

$Z_0 = 360 \ \Omega$

Fig. 6.3 360 Ohm transmission line pair

We know from the original "field strength" equations [1] that the H field is generated by currents flowing in a wire, i.e.,

$$\mathrm{rot}H = i$$

Maxwell [1] taught us that the H field could also be generated by displacement current in free space, i.e.,

$$\mathrm{rot}H = i + \mathrm{d}D / \mathrm{d}t$$

Where $\mathrm{d}D/\mathrm{d}t$ is defined as "displacement current" in free space.

The electromagnetic wave generated by the closely spaced wires and the 360 Ohm transmission line pair were generated by the differential of voltage between the wire pair. The differential in voltage creates an electric flux density "D"

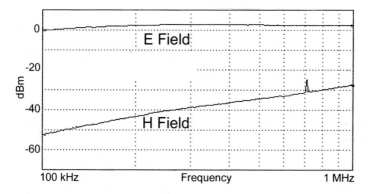

Fig. 6.4 Measured field strengths from loaded electric dipole antenna

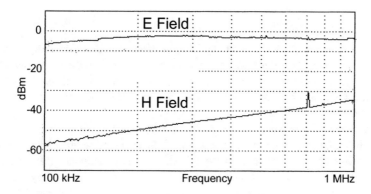

Fig. 6.5 Measured field strengths from 360 Ohm transmission line pair

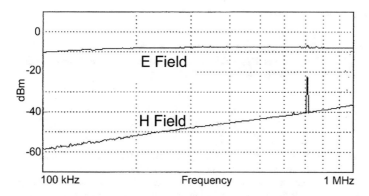

Fig. 6.6 Measured field strengths from closely spaced transmission line pair

Table 6.1 Measured and normalized test data

Radiating element	E field (dB)			H field (dB)			E/H (impedance)	
	Read	Corr factor	Corr data	Read	Corr factor	Corr data	(dB)	Ohms
Frequency—100 kHz								
Electric dipole	0	0	0	−52	−36	−88	88	25×10^3
360 Ohm lines	−6	0	−6	−57	−36	−93	87	22×10^3
Closely spaced wires	−10	0	−10	−59	−36	−95	85	18×10^3
Frequency—1 MHz								
Electric dipole	3	0	3	−28	−36	−64	67	2.2×10^3
360 Ohm lines	−4	0	−4	−34	−36	−70	66	2.0×10^3
Closely spaced wires	−8	0	−8	−37	−36	−73	65	1.8×10^3

between the wire pair. Because the field is generated by a sinusoidal wave, D became dD/dt or displacement current. This displacement current radiates out away from the wire pair, where an H field at right angles to the direction of the displacement current is created. The E field (as described in Chap. 3) is a function of the displacement current.

The electric dipole antenna contains vertically oriented wires containing current flow. This current flow generates a low impedance wave. The horizontal wires contain a differential of voltage, where this difference in voltage generated a high impedance wave. As is illustrated above, the impedance of the wave by the electric dipole antenna is similar to the impedance of the two-wire pair. It can therefore be concluded that the recorded wave is generated by displacement current as a result of the difference in voltage between the horizontal wires.[1]

References

1. Hallen E (1962) Electromagnetic theory. Wiley, New York
2. International Telephone and Telegraph Corp (1956) Reference data for radio engineers, 4th edn. International Telephone and Telegraph Corp., New York

---

[1] Note: The various books on antenna theory [2] inform us that a wave generated by an electric dipole antenna is due to the difference in voltage potential between the ends of the antenna. However, knowing that this voltage generates displacement current which is responsible for the creation of the E and H fields will help design engineers understand how the energy couples to wires and enclosures exists.

Chapter 7
Optimal Use of EMI Gaskets

Electromagnetic Interference (EMI) gaskets are often recommended for use in electrical and electronic systems to (1) reduce the cost of the system and (2) ensure the required shielding is obtained or if the frequency (above 1GHz) mandates the use.

Figure 7.1 illustrates a curve which represents the shielding effectiveness of an enclosure housing with a gap between screws of approximately 6.2 in. The curve can be used (when EMI gaskets are not employed) to predict a level of shielding as a function of the wavelength of the frequency of concern and the screw spacing on a cover.

As is illustrated the peak at approximately 480 MHz will provide 12 dB of shielding. As is shown the level of shielding at 240 MHz (1/2 the frequency) will provide about 40 dB of shielding. The shielding then improves at a rate of 20 dB per decade as the frequency decreases. Therefore, at 24 MHz the shielding will be 60 dB. Using this information, one can estimate the screw spacing required for a given level of shielding at any specific frequency as a function of the wavelength. The example below illustrates how to use the curve of Fig. 7.1.

Example: A computer uses 100 MHz of data on the motherboard. The wavelength is 3 m with a quarter wavelength of 75 cm. With a screw spacing of 75 cm (29.5 in.), the shielding will be 12 dB. At 37.5 cm (14.7 in.) (1/2 the distance), the estimated shielding at 100 MHz will be 40 dB, and at 3.75 cm (1.5 in.) the estimated shielding is 60 dB.

The use of EMI gaskets can be cost–effective by replacing some of the screws and the required nut plates with an EMI gasket where the removal of the screws and nut plates can more than pay for the cost of the gasket.

Figure 7.2 illustrates a cover plate showing a gasketed section.

As is illustrated, the cover bows out between the location of the screws. This bowing is caused by the force of the gasket against the cover. The shielding that is obtained by the gasket (to a large extent) is a function of the shielding obtained halfway between the screws or when the force of the gasket against the cover is minimal.

© Springer Nature Switzerland AG 2020
G. M. Kunkel, *Shielding of Electromagnetic Waves*,
https://doi.org/10.1007/978-3-030-19238-9_7

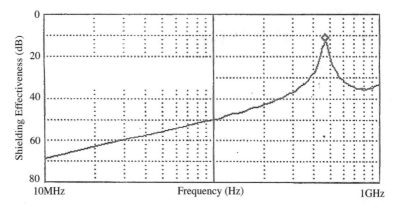

Fig. 7.1 Shielding effectiveness of enclosure with 6.2 in. screw spacing. (The 6.2 in. (15.7 cm) represents ¼ wave length at 480 MHz.)

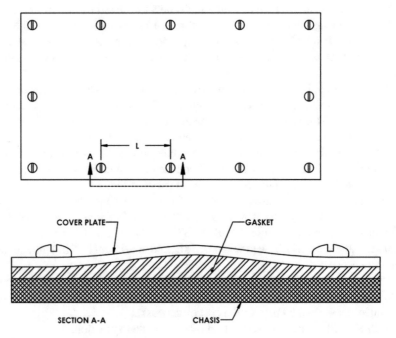

Fig. 7.2 Typical gasketed cover plate

The following are equations [1] that can be used to obtain the maximum screw spacing as a function of the force of the gasket against the cover plate, the diameter (or thickness) of the gasket and the thickness of the plate.

Equations for Predicting Maximum Screw Spacing

$$L = \left[\frac{480\ y\ E\ t^3\ d}{13\ F_1 + 2\ F_2} \right]^{1/4} \quad \text{Fastener Separation (in.)}$$

$$F_0 = \frac{(2F_1 + F_2)\ L}{3} \quad \text{Fastener Force (lb)}$$

where

y = Apparent width of cover flange (in.)
       (Assume 1 in. if not distinct)
t = Thickness of cover edge (in.)
d = Δ deflection of gasket (in.)
F_1 = Minimum force on gasket (lb/in.)
F_2 = Maximum force on gasket (lb/in.)
E = Modulus of Elasticity of cover plate
   = 10^7 for Aluminum (psi)
   = 3×10^7 for Steel (psi)

In using the equations above, the recommended compression at the screw location is 25% with a minimum compression (half way between screws) of 5%. If a gasket is 1/8 (0.125) inches in diameter (or thickness), requires 30 lb/in. to compress it 25%, the maximum force is 30 lb/in. with a minimum force of 6 lb/in. The "d" dimension in the equation is the amount of deflection of the cover or 20% or 0.025 in. As an example, if the cover plate is 0.090 in. thick aluminum using the gasket defined above the maximum screw spacing is:

$$\left[\frac{(480)(1.0)(10^7)(0.090^3)(0.025)}{13(6) + 2(30)} \right]^{1/4} * 0.90 = 4.5 \text{ in.}$$

Note: the 0.90 represents a safety factor.

Using a screw spacing of 4.5 in. the force each screw is to deliver is:

$$\frac{(12 + 30)4.5}{3} = 63 \text{lbs}$$

The shielding offered by the gasket will be consistent with the minimal deflection of a gasket compressed 5% or 0.006 in.

Reference

1. Kunkel G (1980) Mechanical force considerations in the use of EMI-environmental gaskets. IEEE, EMC International Symposium, Baltimore, MD

Chapter 8
Review of Current Test Methods Used to Grade EMI Gaskets by the Industry

A Faraday Cage is a shielded enclosure free of seams and apertures. By definition: when an electromagnetic wave is impinged onto the enclosure, the wave passes the enclosure and does not penetrate the enclosure. When a wave is generated inside the enclosure, the wave does not penetrate the enclosure.

Today's electronic equipment must be designed to (1) operate in the electromagnetic (EM) environment it is to be employed in and (2) ensure that the EM fields generated by the system do not add significantly to the EM environment. This is accomplished by having the equipment housing simulate a "Faraday Cage" (i.e., the electronic system functions properly without adding to the EM environment). The proper selection and use of an EMI gasket to seal the seams of an enclosure is the secret to obtaining a simulated Faraday Cage.

The ability of EMI gaskets to seal the seam of an equipment housing to create a simulated Faraday Cage is a function of the ability to test for the following variables [1]:

1. The conductivity of a gasket when applied to the joint surfaces to be used in actual application.
2. The level of shielding a gasket will provide when the force of the gasket against the joint surface is a minimum.
3. When required, effects moisture and salt fog environments have on the conductivity of the gasket when in contact with the joint surfaces in actual use.

There are presently two (2) methods employed by industry to test EMI gaskets. These are Shielding Effectiveness (as contained in MIL-DTL-83528) and Transfer Impedance (per SAE ARP-1705).

Transfer Impedance Test Methods have the ability to test all the variables of concern to an accuracy of ± 2 dB over the frequency range of 10 kHz to 18 GHz and discriminate between the gaskets on the market over a dynamic range of greater than 120 dB.

© Springer Nature Switzerland AG 2020
G. M. Kunkel, *Shielding of Electromagnetic Waves*,
https://doi.org/10.1007/978-3-030-19238-9_8

Shielding Effectiveness test method is known to (1) provide highly inflated levels of shielding (a newspaper has produced over 100 dB of shielding at 2 GHz); (2) does not test for the critical variables; (3) provides the relative level of conductivity of a gasket over the dynamic range of 30 dB (versus 120 dB for transfer impedance); and (4) cannot be used to predict a relative level of shielding at any given frequency.

Transfer Impedance Testing

The Transfer Impedance testing used by industry is per SAE ARP-1705 Rev. C. The original ARP was prepared by the electromagnetic pulse (EMP) engineering community to test EMI gaskets for compliance to the EMP requirements. The original document has been upgraded to include the needs of the EMC engineering community. These revisions have made it possible to:

1. Measure the conductivity of a gasket against any of the structural materials and finishes used by industry.
2. Measure the conductivity of a gasket when used against any of the structural material and finishes as a function of the force against the joint surfaces.
3. Measure the effects moisture and salt fog environments have on the gasketed application.

Figures 8.1 and 8.2 illustrate significant information to a design engineer in the selection of a gasket for a specific application [1]. As is illustrated in Fig. 8.1, the ability of the EMI gaskets on the market to seal the seams of an enclosure varies by more than 120 dB. A significant value to the design engineering community is that if a given gasket fails to provide the required shielding, a more conductive gasket can easily be selected.

Figure 8.2 illustrates the importance of plating the joint surfaces have on the shielding a gasket can provide. Also illustrated is the effect 336 h of humidity soak has on the shielding as a function of the plating [1].

Shielding Effectiveness Testing

The shielding effectiveness testing as performed by industry is contained in MIL-DTL-83528 Rev. G. The test is a modified MIL-STD-285 (IEEE 299) test [1]. The modifications are:

1. The transmitting antenna is aimed at the center of a large (26″ by 26″) 3/8 in. thick aluminum plate instead of at the gasket under test, and.
2. The receiving antenna is aimed at the center of the large aluminum plate instead of recording the maximum field strength radiating from the gasketed area.

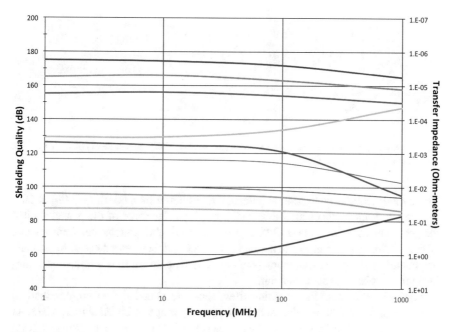

Fig. 8.1 Transfer impedance test data of various EMI gaskets on the market against Tin Plated Surfaces

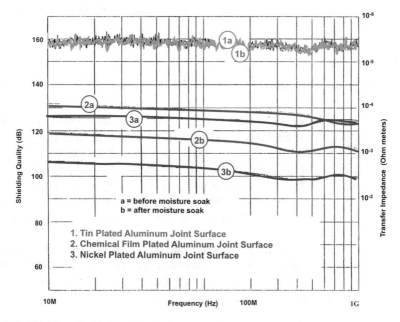

Fig. 8.2 Shielding of a Tin/Lead Plated Spiral Type gasket against tin, chemical film, and nickel-plated aluminum before and after being subjected to a 336 h moisture soak

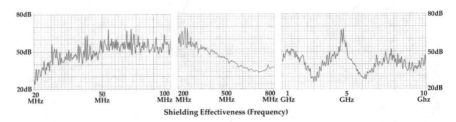

Fig. 8.3 Shielding effectiveness test data performed per MIL-DTL-83528 Rev G on a 1/8 in. thick (non-conductive) phenolic gasket

These modifications can produce inflated data of better than 60 dB (see Fig. 8.3).

The test joint surfaces are brass and aluminum. The force of the gasket against the joint surfaces is uncontrolled and can exceed 400 pounds per inch (where 1.0 pound/in. can be realized in actual use). The test data cannot be used to estimate the shielding at a given frequency (see Fig. 8.3). The test time is hours versus minutes for transfer impedance testing.

Figure 8.3 illustrates the shielding effectiveness test results[1] of a (non-conductive) phenolic gasket 0.125 in. thick over the frequency range of 20 MHz to 10 GHz. As is illustrated the shielding varies between 68 dB and 24 dB where the expected level of shielding approaches four (4) dB over the frequency range. This renders some of the data inflated by more than 60 dB. This type of data illustrated on Fig. 8.3 is expected to be obtained from all the gaskets tested to the standard. As is observed, such data is worthless to a design engineer in the selection of a gasket for a specific application.

Summary

Transfer Impedance test data provides a very accurate (±2 dB) level of the conductivity of an EMI gasket over the frequency range of 10 kHz to 18 GHz against any of the structural materials and finishes used by the electrical and electronic industry. This includes force of a gasket against the joint seam material. The ability to withstand moisture and salt fog environments can also be obtained when required. This conductive data can easily be converted to fairly accurate levels of shielding. In the event a selected gasket does not comply as required, a suitable gasket can readably be obtained.

Shielding effectiveness data is highly inflated and erratic. Specific levels of shielding at a given frequency are not obtainable. A general "good/better/best" over a dynamic range of 30 dB is provided. However, this good/better/best can vary as a function of the structural material and finish a gasket is to be used against as well as

---

[1] Note: The testing was performed at DNB Engineering in Fullerton California in their anechoic test chamber modified to test to MIL-DTL-83528.

the actual force to be applied. If the selected gasket fails to comply with a required Specification, there is no guarantee a solution can be found using the test data.

Conclusion

The shielding effectiveness test method (used by the major manufacturers of EMI gaskets to grade their products) produces highly flawed and erratic data. As such, the data cannot be used by the design engineering community in the selection of an EMI gasket for an actual application. Transfer impedance test data provides all the information required.

It appears unlikely that the major manufacturers of EMI gaskets will supply their customers with transfer impedance test data on their products (unless they are forced to by their customers). However, if a company is interested in obtaining transfer impedance test data, it can be obtained at a fairly low cost. The equipment needed is a spectrum (or network) analyzer and the transfer impedance test fixtures. The spectrum (or network) analyzer used to test for system EMC compliance is often already available in an in-house testing facility. The transfer impedance test fixtures are relatively inexpensive where a test will take 10–15 min to complete.

Reference

1. Kunkel G (2018) Testing of EMI gaskets. Evaluation Engineering Magazine

Chapter 9
Silver Elastomeric Specification MIL-DTL-83528

MIL-DTL-83528 Rev. G is a specification covering Silver Elastomeric EMI gaskets "approved for use by all departments and agencies of the department of defense" [1].

The document is a general specification to be used for purchasing silver elastomeric EMI gaskets for the purpose of providing environmental and shielding protection for Department of Defense (DoD) weapon systems. The silver elastomeric gaskets referenced in this specification can possess two (2) significant problems for the DoD. These problems are:

1. The gaskets can possess a relatively short shelf life (the ability to shield a weapon system from the effects of an electromagnetic wave can be compromised while sitting on the shelf.)
2. The gaskets can possess a relatively short useful life. The silver content of the gasket material when exposed to moisture or salt fog environments will create a galvanic corrosion between the gasket and the structural surface of a weapon system [2]. This galvanic corrosion will highly compromise the ability of the gaskets to provide the required shielding.

A full critique of MIL-DTL-83528 G verifies that the problems exist, and is contained in Appendix A.

References

1. MIL-DTL-83528 Rev. G (2017) Gasketing material, conductive, shielding gasket, electronic, elastomer EMI/RFI general specification, US Department of Defense
2. SAE ARP-1481 Rev. A (2004) Corrosive controlled and electrical conductivity in enclosure design, SAE International

© Springer Nature Switzerland AG 2020
G. M. Kunkel, *Shielding of Electromagnetic Waves*,
https://doi.org/10.1007/978-3-030-19238-9_9

Chapter 10
Shielding Effectiveness Theory of Shielding

The hypothesis upon which the presently accepted shielding effectiveness theory of shielding is based is as follows [1]:

1. Wave theory represents the actual phenomenon associated with a wave traveling on a transmission line.
2. The penetration of a wave into and through a shielding barrier is analogous to a wave on a transmission line as predicted by wave theory.

The theory is based on the following assumptions:

1. The wave is a vector consisting of an E and H field.
2. When the wave is impinged on a barrier, the E field is reflected where the value of the reflection is provided by wave theory reflected loss equation.
3. The wave is attenuated by skin effect as the wave penetrates the barrier.
4. When the wave strikes the exit surface of the barrier, the H field is reflected due to the difference in the impedance of the barrier and free space.
5. The impedance of the wave as it leaves the barrier is the same as Free Space or 377 ohms.

The results of the analysis as contained herein illustrate that for wave theory to yield accurate results for a transmission line, specific boundary conditions must be met. Based on the results, it can be concluded that wave theory does not represent a wave on a transmission line and that wave theory is only a theory. Therefore, the hypothesis upon which the currently accepted "Theory of Shielding" is based does not exist. The theory can, therefore, be assumed to be inaccurate.

© Springer Nature Switzerland AG 2020   39
G. M. Kunkel, *Shielding of Electromagnetic Waves*,
https://doi.org/10.1007/978-3-030-19238-9_10

Analysis: Circuit Theory Versus Wave Theory [1]

The analysis contained herein is to compare the results of calculating the current through the load of various circuits using wave theory and circuit theory equations.

Example circuit (Fig. 10.1):

Let Z_L and Z_S equal the following:

(a) $Z_S = 50\ \Omega$, $Z_L = 50\ \Omega$
(b) $Z_S = 50\ \Omega$, $Z_L = 1\ \Omega$
(c) $Z_S = 50\ \Omega$, $Z_L = 2500\ \Omega$
(d) $Z_S = 50\ \Omega$, $Z_L = 1 + j49.99\ \Omega$
(e) $Z_S = 1 - j49.99\ \Omega$, $Z_L = 1 + j49.99\ \Omega$

Using circuit theory: $\left(I = \dfrac{E}{Z_S + Z_L} \right)$

(a) Z_S & $Z_L = 50\ \Omega$, $I = 1.0$ Amps
(b) $Z_S = 50\ \Omega$, $Z_L = 1\ \Omega$, $I = 1.9608$ Amps
(c) $Z_S = 50\ \Omega$, $Z_L = 2500\ \Omega$, $I = 0.0392$ Amps
(d) $Z_S = 50\ \Omega$, $Z_L = 1 + j49.99\ \Omega$, $I = 1.4105$ Amps
(e) $Z_S = 1 - j49.99\ \Omega$, $Z_L = 1 + j49.99\ \Omega$, $I = 50.0$ Amps

Using wave theory:

$$\text{Reflected loss}(\text{RL}) = (K+1)^2 / 4(K)$$

where $K = \dfrac{Z_O}{Z_{\text{Load}}}$

(a) Z_S & $Z_L = 50\ \Omega$, $\text{RL} = \dfrac{(1+1)^2}{4(1)} = 1.0$

(b) $Z_S = 50\ \Omega$, $Z_L = 1\ \Omega$, $\text{RL} = \dfrac{(50+1)^2}{4(50)} = 13.005$

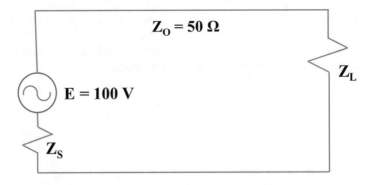

Fig. 10.1 Typical circuit

(c) $Z_S = 50\ \Omega$, $Z_L = 2500\ \Omega$, $RL = \dfrac{(0.02+1)^2}{4(0.02)} = 13.005$

(d) $Z_S = 50\ \Omega$, $Z_L = 1 + j49.99\ \Omega$, $RL = \dfrac{(1+1)^2}{4(1)} = 1.0$

(e) $Z_S = 1 - j49.99\ \Omega$, $Z_L = 1 + j49.99\ \Omega$, $RL = \dfrac{(1+1)^2}{4(1)} = 1.0$

The reflective coefficient (RC)[1] is as follows:

(a) Z_S & $Z_L = 50\ \Omega$, RC = 0 or 0% Reflected
(b) $Z_S = 50\ \Omega$, $Z_L = 1\ \Omega$, RC = 0.923106 or 92.3106% Reflected
(c) $Z_S = 50\ \Omega$, $Z_L = 2500\ \Omega$, RC = 0.923106 or 92.3106% Reflected
(d) $Z_S = 50\ \Omega$, $Z_L = 1 + j49.99\ \Omega$, RC = 0 or 0% Reflected
(e) $Z_S = 1 - j49.99\ \Omega$, $Z_L = 1 + j49.99\ \Omega$, RC = 0 or 0% Reflected

When:

(a) Z_S & $Z_L = 50\ \Omega$
(d) $Z_S = 50\ \Omega$, $Z_L = 1 + j49.99\ \Omega$
(e) $Z_S = 1 - j49.99\ \Omega$, $Z_L = 1 + j49.99\ \Omega$

The reflected power is zero (0)

(b) $Z_S = 50\ \Omega$, $Z_L = 1\ \Omega$
(c) $Z_S = 50\ \Omega$, $Z_L = 2500\ \Omega$

The reflected power is: (0.923106) (50) = 46.1553 watts

Therefore, the power absorbed by the load impedances when $Z_S = 50\ \Omega$ and $Z_L = 1.0$ or $Z_L = 2500\ \Omega$ is:

50 – 46.1553 = 3.8447 watts (50 watts is the power delivered to the load when $Z_L = 50\ \Omega$)

The current through the load impedance using wave theory is:

1. Z_S & $Z_L = 50\ \Omega$, $I = 1.0$ Amps
   (d) $Z_S = 50\ \Omega$, $Z_L = 1 + j49.99\ \Omega$, $I = 1.0$ Amps
   (e) $Z_S = 1 - j49.99\ \Omega$, $Z_L = 1 + j49.99\ \Omega$, $I = 1.0$ Amps

2. $Z_S = 50\ \Omega$, $Z_L = 1\ \Omega$, $I(1.0) = 3.8447$
   $I = 1.9608$ Amps

3. $Z_S = 50\ \Omega$, $Z_L = 2500\ \Omega$
   $I^2 = 3.8447/2500$
   $I = 0.0392$ Amps

Table 10.1 illustrates the results of the analysis.

---

[1] Note: 1/RL yields percent or power absorbed by load
RC = 1-(1/RL) yields percent of reflected power

Table 10.1 Results of analysis (amperes)

Source impedance (Z_S)	Load impedance (Z_L)	Results of analysis	
		Circuit theory	Wave theory
50	50	1.0	1.0
50	1.0	1.9608	1.9608
50	2500	0.0392	0.0392
50	$1 + j49.99$	1.4105	1.0
$1 - j49.99$	$1 + j49.99$	50.0	1.0

Summary of Results

The analysis as contained herein illustrates that specific boundary conditions must be met for wave theory to yield accurate results. These conditions are:

1. The circuit must be a simple circuit consisting of a power source, source impedance, and load impedance.
2. The impedance must be resistive.
3. The transmission line must be short enough that the load impedance, as viewed from the source, is resistive.

These boundary conditions highly restrict the use of wave theory and illustrate that wave theory is only a theory and does not generally represent a wave on a transmission line. This renders the hypothesis upon which "Shielding Theory" [2] (as currently accepted by the electrical/electronic industry) inaccurate. Additionally, shielding effectiveness tests have been performed on space blanket material (having an impedance of 2.0 ohms). The results of the tests illustrate no detected H field reflection. Additionally, the impedance of the wave leaving the "space blanket" was the same as the space blanket.

References

1. Kunkel G (2018) Testing of EMI gaskets. Evaluation Engineering Magazine
2. Frederick Research Corp (1962) Handbook of radio frequency interference, vol 3 (Methods of Electromagnetic Interference Suppression). Frederick Research Corp., Wheaton, MD

Chapter 11
Shielded Wires and Cables

Shielded wires and cables are used within electrical and electronic systems to reduce (to within tolerable limits) electromagnetic interference (EMI) within a system.

The basic shielding supplied by a shielded wire or cable is 60 dB [1]. This level of shielding is obtained by a single layer of braid. This basic level tends to disappear (due to standing waves on the shield as a result of transmission line effects) as a function of the length of the wire or cable and the wavelength of the frequency of interest. When the length of the wire or cable is equal to ½ of the wavelength of the frequency of concern (given the shield is terminated at both ends), the shielding approaches zero. When the shield is referenced to ground at only one end, the shielding tends to disappear when the length of the wire or cable is equal to ¼ wavelength.

All shielded wires within a system are used to (1) contain interference generated by the signals being transmitted within the wire and (2) protect the circuits associated with the wire or wires from EMI. The termination of the shield is critical. The shield in most cases is reference to ground at only one end (to eliminate capacitive coupling from wire to wire). When a shield is used to keep the interference contained in the wire from radiating out, the termination should be at the source of the signal. When the shield is used to protect the circuits associated with the wire from interference, analysis should be conducted to determine the proper location (the current that is capacitively coupled to the shield must return to its source using the ground reference system). The area of concern is the ground reference system. The base emitter junction of a transistor contains a diode that is triggered when the differential voltage exceeds 0.7 V. When the ground reference voltage differential between a transmitting and receiving circuit approaches this 0.7 of a volt ground reference interference can occur (this is often referred to as ground loop interference). When the circuits of the system are held to an equal potential using the chassis of a system or subsystem, a ground plane close to the chassis reference should represent the optimal position to reference the shield.

Wire coupling consists of inductive as well as capacitive coupling. Inductive coupling becomes a problem when the current in a wire is fairly high. In the case of

© Springer Nature Switzerland AG 2020
G. M. Kunkel, *Shielding of Electromagnetic Waves*,
https://doi.org/10.1007/978-3-030-19238-9_11

inductive coupling, the shield or shields should be terminated to ground at both ends. It is critical that the resistance between the two ends is held to a minimal. To accomplish this, the system chassis should be used to the greatest extent possible.

When cables are used internally to a system or subsystem, it can be assumed that the interference of concern is both capacitive and inductive. In the use of a shielded cable, it is critical that the impedance between the two ends be held to a minimum. If a system employs a single point ground, referencing a shield or cable to ground at both ends is difficult if not impossible.

Shielded cables tying subsystems together to form a large system are often employed. The shields on these cables can consist of a single layer of braid. Two layers of braid will add approximately 6 dB to the shielding. When the two layers of braids are separated by a nonconductive dielectric film, the shielding can be improved by as much as 40 dB. Many authors have proposed a shielded cable consisting of the inner braid reference to ground at only one end with an over braid separated by a nonconductive film referenced to ground at both ends. A shielded cable constructed in this method will provide a high level of shielding while illuminating the possibility of the cable providing zero shielding. The preferred method of referencing the shield to ground potential is with the use of a back shell that is constructed in such a method as to provide a positive ground reference. However, such cables are often referenced to ground with a wire that is brought through the interface connector and reference to ground inside the system. This type of referencing the shield has two problems associated with it. If the wire is not held to a very short length, the inductance and resistance of the wire can minimize or eliminate the effect of the shield. The second problem is that currents coupled to the shield are brought into the system. These currents can create a ground reference problem.

When back shells are used to reference a cable to ground potential, the back shell is designed for a specific cable as well as a specific connector. The connector consists of two parts. These are a connector that is attached to the back shell and a receptacle that references the cable to an equipment chassis. There are two types of adapters. These are a back mount and a front mount adapter. The back mount is referenced to the chassis ground on the outside of the equipment housing. The advantage of a back-mount receptacle is that currents coupled to the cable are referenced to ground outside of the equipment chassis. The disadvantage is that the wires internal to the system must be tied to the receptacle with the internal equipment housed inside the chassis. This represents a manufacturing problem. A front mount receptacle is mounted inside the equipment enclosure. The advantage is that the wires leading to the front mount adapter can be attached to the adapter off-line. The disadvantage is that the currents coupled to the cable are brought into the equipment chassis. High currents such as those associated with EMP waves and lightning can cause disruption and possibly destruction of circuits.

A coaxial cable is a special type of shielded cable. These cables consist of a single conductive wire where the characteristic impedance of the cable is 50 ohms. The cables are always terminated into a connector or adapter that maintains the 50 ohm characteristic impedance. The sending and receiving circuits reference to the cable are 50 ohms. Maintaining the 50 ohms throughout the entire system reduces (to the greatest extent possible) standing waves and reflections in the cable. The current flowing in the wire from the transmitting circuit generates an H field that creates a back EMF on the shield. The voltage which is opposite to the direction of the current in the wire causes the current returning to the source to flow on the shield. The result is that the H field on the shield is equal to and opposite the H field created by the current in the wire where the H field generated by the cable approaches zero. The shielding on these cables can vary between 60 dB having a single layer of braid to well over 100 dB. There are numerous types of coaxial cables, where the different types are a function of the maximum frequency the cables are designed to handle. The level of shielding can vary significantly and is a function of the type of shield employed on the cable.

Reference

1. White DRJ (1973) Electromagnetic interference and compatibility, vol 3 (EMI Control Methods and Techniques). Don White Consultants Inc., Germantown, MD

Chapter 12
Shielding Air Vent Materials

Shielded cabinets, enclosures, and many electrical and electronic systems require the passage of air along with maintaining the required level of shielding. The type of material used to obtain the required airflow and shielding is a function of the required amount of airflow, the level of shielding, and the frequency range for the required shielding. The different materials are (1) conductive metal screening; (2) perforated metal; and (3) conductive honeycomb panels.

1. Conductive metal screening. The shielding associated with the conductive metal screening can reach 60 dB at 1 GHz and is a function of the type and dimension of the wire and the tightness of the weave. The airflow is highly restricted.
2. Perforated metal panels. The level of shielding and airflow properties of the perforated metal panels are a function of the size of the holes and percentage of open area. Both the shielding and airflow should be better than the conductive metal screening.

The use of conductive metal screening and perforated metal materials can be cost-effective but requires extensive engineering. The level of shielding and airflow must be tested and analyzed for a specific application. The electromagnetic (EM) bonding of the material to the chassis must be designed in house. As such the use is only recommended for large production runs.

Honeycomb Panels

The airflow properties of the honeycomb panels are excellent. The level of shielding varies significantly and is a function of the method used to electromagnetically bond the metal foil that forms the cells together. The methods currently in use are (1) soldering; (2) spot welding, and (3) nonconductive epoxy. The soldering is used on brass and iron (or steel) foil and represents the highest level of shielding and is effective to 10 GHz. The soldered brass or iron honeycomb panels are basically

© Springer Nature Switzerland AG 2020
G. M. Kunkel, *Shielding of Electromagnetic Waves*,
https://doi.org/10.1007/978-3-030-19238-9_12

1/8-in. cells by ¼ and ½ in. thick. The shielding provided by these panels at 1.0 GHz is 108 dB and 128 dB, respectively. The soldered brass and iron panels are also offered with ¼ in. cells by 1.0 in. thick. These panels provide shielding levels of over 140 dB and are used in shielded enclosures and cabinets designed to provide 120 dB of shielding at 10.0 GHz.

The spot welding of honeycomb panels is performed on stainless steel panels. These panels typically come in 1/8-in. cell by ¼ and ½ in. thick. The shielding at 1.0 GHz for a ¼ in. thick panel with a single spot weld between foils is approximately 60 dB. Additional spot welds per cell will increase the level of shielding. The ½ in. thick panels will provide about 6 dB above that of the ¼ in. thick panels.

Aluminum Honeycomb Panels

The cells of the aluminum honeycomb panels are bonded together with nonconductive epoxy. Hexcel holds the original patent on the manufacturing process. In the manufacture of the panels, the manufacturer makes a loaf of honeycomb 4 ft. by 8 ft. by 36 inches high. They then slice the loaf into 4 by 8 foot panels to any required thickness. In the process of slicing the panels, incidental contact points are created that bridge the epoxy gap. The shielding which is obtained from the panels is a function of how many contact points that bridge the epoxy gap are created. The actual measured shielding of the panels varies between 30 dB and 60 dB for a panel with 1/8-in. cells by ¼ in. thick. A ½ in. thick panel will add about 6 dB. Tin plating is often recommended. However, the caustic-etch used to remove the oxide coating from the aluminum panels required for plating can remove the incidental contact points. The loss of the contact points can result in a significant loss of shielding.

Figure 12.1 illustrates the measured level of shielding of ¼ in. thick panel. Figure 12.2 illustrates a magnified end view illustrating the existence of the incidental contact points [1].

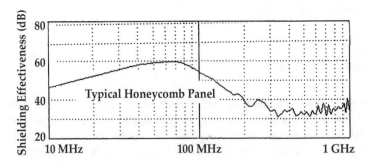

Fig. 12.1 Typical shielding of aluminum panel

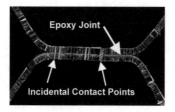

Fig. 12.2 End view illustrating contact points

There is a patented "blending" process for aluminum honeycomb panels that results in a positive bridging of the epoxy gap forcing one of the foils to bridge the gap making contact with the adjacent foil. Figure 12.3 illustrates the level of shielding obtained from a 1/8-in. cell by ¼ in. thick panel before and after applying the blending process [1].

Figure 12.4 illustrates the magnified end view of a panel after being blended. As is illustrated the epoxy gap is totally bridged.

All vendors which supply aluminum honeycomb filters provide a double panel. The double panel provided by the vendors using the panels as delivered from the manufacture typically appreciates a 6 dB improvement. The double panel supplied by the vendor having the patented blending process (using patented processes) provides an added 40 dB of shielding resulting in shielding level of 108 dB at 1.0 GHz for two 1/8-in.-thick panels and 128 dB for two ¼ in. panels.

Honeycomb Panel Airflow Properties

The airflow properties of the honeycomb panels in terms of the "static Air Pressure Drop" versus air speed are a function of the cell size, length of the cells (width of the panels), and thickness of the foil. The double panel adds to the static air pressure due to the added edge of the additional panel. Note! The brass (or iron) honeycomb panels are made from thicker foil material than the aluminum. This results in higher resistance where the brass (or iron) panels exhibits similar airflow properties with the double aluminum panels.

Figure 12.5 illustrates airflow properties of the various honeycomb panels [1]. Note: values of the stainless-steel panels have not been made available to the author, but should be consistent with the soldered honeycomb materials.

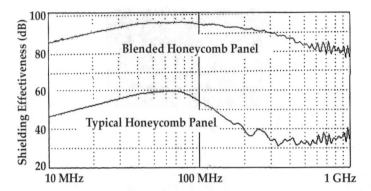

Fig. 12.3 Typical aluminum honeycomb panel before and after blending

Fig. 12.4 End views of blended panel

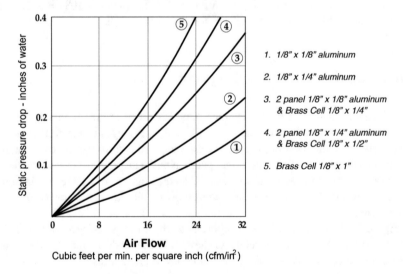

Fig. 12.5 Airflow properties of honeycomb panels

Reference

1. Spira Manufacturing Corp (2018) EMI gaskets and shielding products, catalog and design guide. Spira Manufacturing Corp, San Fernando, CA

Chapter 13
Shielding Effectiveness Testing of Space Blanket

Shielding effectiveness testing of a 1.4 ohms per square (impedance of 2.0 ohms) space blanket has been performed, where the results of the test were recorded. This recorded data illustrated that the theory of shielding as accepted using the eq. $SE = R + A + B$ yields inaccurate results.

1. A test was performed [1] at DNB Engineering in Fullerton California in one of their shielded enclosure rooms that was modified to test Electromagnetic Interference (EMI) gaskets per the requirements of MIL-DTL-83528 [1]. The testing consisted of measuring the E and H fields from both an electric and magnetic dipole antenna. The frequency range was from 100 kHz to 10 MHz.
2. A demonstration was performed at the 1999 IEEE EMC symposium in Seattle [2]. The testing was performed using a small shielded enclosure. An electric dipole antenna was placed inside the enclosure, where a space blanket material was used at the face of the enclosure to illustrate the shielding of the space blanket. The shielding which was illustrated showed the attenuation of both the E and H fields. The frequency range was from 100 kHz to 1 MHz.

The shielding effectiveness test performed at DNB Engineering was conducted in a shielded room modified to perform testing in accordance with MIL–DTL–83528. The test was conducted through the 24 × 24″ hole in the wall. The radiating antennas were (1) electric dipole antenna 4 cm from the test sample and (2) a magnetic dipole antenna 30 cm from the test sample. The test consisted of measuring the E and H fields from each of the antennas before and after placing the test sample onto the 24 × 24″ hole in the wall of the enclosure. The significant findings were that there was not an H field shielding from either of the sources of the wave.

Figures 13.1 and 13.2 illustrate the recorded H field from the electric dipole antenna over the frequency range of 100 kHz through 1.0 MHz (The recorded H field data over the frequency range of 1.0 MHz to 10 MHz and the recorded data from the magnetic dipole antenna is similar).

© Springer Nature Switzerland AG 2020
G. M. Kunkel, *Shielding of Electromagnetic Waves*,
https://doi.org/10.1007/978-3-030-19238-9_13

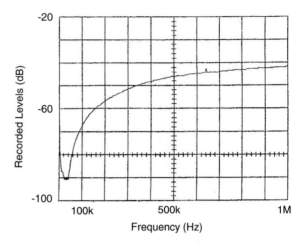

Fig. 13.1 *H* field test data from electric dipole antenna *without* test sample

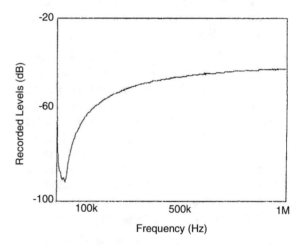

Fig. 13.2 *H* field test data from electric dipole antenna *with* test sample

The results of the *E* field shielding as recorded from the test at DNB Engineering were compared to the attenuation as predicted by the shielding effectiveness theory consisting of SE = *R* + *A* + *B*.

Table 13.1 illustrates the results of the shielding effectiveness analysis using the test conditions employed at DNB Engineering.

The results of the *E* field shielding effectiveness test performed at DNB Engineering are compared with the analysis of Table 13.1. Figures 13.3 and 13.4 illustrate the recorded shielding effectiveness to the *E* field as compared with the predicted shielding effectiveness.

As is illustrated in Figs. 13.3 and 13.4 there is a relationship between the recorded *E* field and the shielding effectiveness analysis. However, the level of predicted shielding varied from that of recorded by a fairly significant difference.

Table 13.1 Shielding effectiveness analysis of test conditions used at DNB Engineering

Antenna source	Z_w	Z_b	K	δ	$e^{-2t/\delta}$	$\left(\dfrac{k-1}{k+1}\right)^2$	R dB	B dB	A dB	SE dB
@ 100 kHz										
High Z	9×10^5	0.0001	6×10^9	0.0003	0.9998	1.0000	183.6	−76.4	0.0	107
Low Z	0.158	0.0001	1058	0.0003	0.9998	0.9962	48.4	−48.1	0.0	0
@ 1 MHz										
High Z	90,000	0.0005	190×10^6	0.0001	0.9995	1.0000	153.6	−66.4	0.0	87
Low Z	1.0000	0.0005	2120	0.0001	0.9995	0.9981	54.5	−52.5	0.0	2
@ 10 MHz										
High Z	9000	0.0015	6.03×106	2.67×10^5	0.9905	1.0000	123.6	−56.4	0.0	67
Low Z	15.79	0.0015	16×10^3	2.67×10^5	0.9905	0.9996	68.5	−54.5	0.0	14

High Z represents the electric dipole antenna. Low Z represents the magnetic dipole antenna.

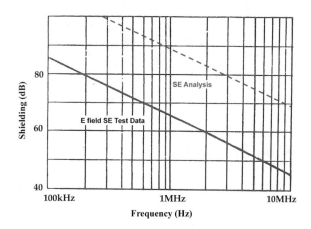

Fig. 13.3 E field shielding effectiveness using electric dipole antenna test data versus shielding effectiveness (SE) analysis

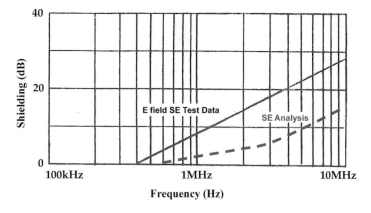

Fig. 13.4 E field shielding effectiveness test data using magnetic dipole antenna versus shielding effectiveness (SE) analysis

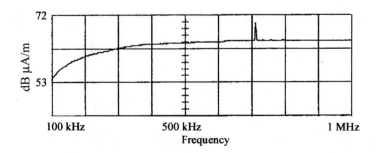

Fig. 13.5 *H* field test data with face of shielded enclosure open

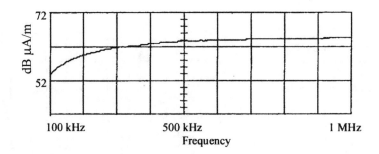

Fig. 13.6 *H* field test data with shield attached to face of enclosure

The demonstration was performed at the 1999 IEEE EMC symposium in Seattle [2]. A small enclosure measuring 8 in. wide by 12 in. high by 4 in. deep contained an electric dipole antenna. The shielding to the *E* and *H* field was demonstrated, where no *H* field shielding was recorded. Figures 13.5 and 13.6 illustrate the recorded *H* field before and after placing a space blanket on the face of the enclosure.

The *E* field recorded data before and after placing the space blanket on the face of the enclosure is illustrated in Fig. 13.7.

The impedance of the wave leaving the barrier was the same as the impedance of the barrier as illustrated in Table 13.2.

Fig. 13.7 *E* field test data

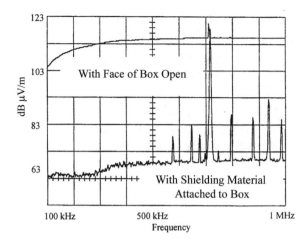

Table 13.2 Impedance of wave leaving blanket

Frequency (Hz)	100 k	500 k	1.0 M
E field (dBμV)	63	69	70
H field (dBμA)	55	64	65
Z_W (*E/H*) dBΩ	8	5	5
Z_W ohms	2.5	1.8	1.8

Summary

The use of a 2.0 ohm space blanket for performing a shielding effectiveness test of both the E and H fields from electric and magnetic dipole antennas produces extremely valuable information.

The presently accepted theory of shielding associated with the eq. $SE = R + A + B$ is based on the following assumptions:

1. The wave is a vector consists of *E* and *H* fields.
2. When the wave strikes a barrier, there is an *E* field reflection.
3. The wave inside the barrier is attenuated by skin effect as the wave penetrates the barrier.
4. When the wave reaches the exit surface, there is an *H* field reflection equal in value to that of the *E* field reflection at the incident surface of the barrier.
5. The impedance of the wave that penetrates the barrier is equal to the impedance of free space or 377 ohms.

The recorded shielding effectiveness data using a space blanket as the test sample yielded the following information:

1. There is not a *H* field reflection at the exit surface of a barrier.
2. The impedance of the wave as it leaves the barrier is equal to the impedance of the barrier.

These results illustrate that the accepted theory of shielding yields inaccurate results. There was a relationship between the *E* field results obtained at DNB Engineering; however, there was a significant difference of the magnitude.

References

1. Broaddus Al, Kunkel G (1992) Shielding effectiveness testing of 1.0 Ohm aluminized Mylar, IEEE, EMC International Symposium, Anaheim, CA
2. Kunkel G (1999) Penetration of electromagnetic fields through shielding barrier material (A demonstration). IEEE/EMC International Symposium, Seattle, WA

Chapter 14
Test Methods for Testing EMI Gaskets: A Review of IEEE 1302

The IEEE 1302 specification [1] references test methods available to test EMI gaskets. The document is intended to be a guide in the selection of the appropriate test method to determine the level of electromagnetic shielding provided in various applications.

During World War II, the Department of Defense (DoD) purchased shielding enclosures to house their electrical and electronic equipment. Steel wool was initially used to caulk the seams of the enclosures when the enclosures failed to meet the shielding requirements. The steel wool rusted out almost immediately and the DoD turned to a company (Metal Textiles Corporation) to obtain woven wire to caulk the seams. Metal Textiles Corporation changed their name to Metex. Several of the employees left and formed Tecknit. Both companies used MIL-STD-285 to grade their products [2]. This consisted of (1) obtaining a reference level in free space and (2) taking a second measurement with a gasketed cover attached to a hole in the wall of an enclosure.

In the performance of the testing, the transmitting antenna was inside the enclosure and the receiving antenna outside. The difference in the two measurements was called the shielding effectiveness of the gasket. It eventually became apparent that this test method did not provide accurate measurements and Tecknit and Metex were forced to include a new reference level in the test data. This new reference level was taken with the transmitting antenna inside the enclosure with the gasketed cover removed from the enclosure. Figure 14.1 illustrates the new data. As can be seen at 50 MHz, the original shielding was claimed to be 90 dB. With the new reference level, it is apparent that the shielding was only 10 dB.

A new standard SAE ARP-1173 was generated to supply industry with a test method that would provide more accurate data [3]. This test method consisted of attaching a fairly large shielded box to a wall inside a shielded enclosure. A cover plate was to be attached to the front of the box. Two tests were to be taken. These consisted of (1) the shielding effectiveness of a gasket under test and (2) shielding improvement of the gasket. The reference for the shielding effectiveness test was taken with the cover plate held in place with nylon screws where nonconductive

© Springer Nature Switzerland AG 2020

G. M. Kunkel, *Shielding of Electromagnetic Waves*,

https://doi.org/10.1007/978-3-030-19238-9_14

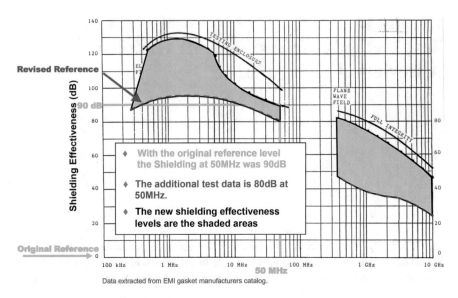

Fig. 14.1 Typical shielding effectiveness test data (extracted from a Tecknit catalog)

washers were placed between the shielded box and the cover plate to provide a gap between the cover plate and the box.

A gasket under test was then placed between the cover plate and the shielded box with a second measurement taken. The difference in the two measurements was called the shielding effectiveness of the gasket. The reference for the shielding improvement test was taken with the cover plate attached to the shielded box. A secondary measurement was taken with the gasket placed between the cover plate and shielded box. The difference in the two measurements was called the shielding improvement of the gasket. These two tests proved to provide very small levels of shielding for the manufacturers of EMI gaskets. As a result, the manufacturers refused to use the test to market their products.

In the late 1950s, the silver elastomeric gasket was introduced to the marketplace. The gaskets proved to provide lower levels of shielding than the wire mesh gaskets. The gasket material was also much more expensive due to the silver content in the gaskets. To improve the marketing ability of the silver elastomeric gasket, the company designed a new type of shielding effectiveness test. This test was classified as a modified MIL-STD-285 test. The modifications were:

1. Aiming the transmitting antenna at a large 26″ × 26″ aluminum plate instead of at the gasketed seam under test.
2. The receiving antenna was aimed at the center of the large 26″ × 26″ aluminum plate instead of measuring the maximum field strength radiating from the gasketed seam.

This new test method results showed high levels of shielding of more than 60 dB. The gasket material was marketed as providing a light weight gasket possessing an environmental seal along with extremely high levels of shielding. These properties were highly desirable by the Air Force where the Air Force designed the silver elastomeric products into a large number of their weapon systems. When the company that developed and manufactured the gasket lost its patent rights, the company convinced the Air Force to prepare a standard around the product to ensure that other companies manufacturing a silver elastomeric gasket would provide gaskets as good as or better than those produced by the original company. The standard which was produced was MIL-G-83528 [4]. The procedure was written by the company that developed and manufactured the original silver elastomeric gaskets. The standard contained the shielding effectiveness test that they had designed to test the products. The original standard was superseded by MIL-DTL-83528 C. Revisions D through G were prepared and monitored by the custodian of the standard but were provided by the various manufacturers of the elastomeric gasket.

In 1962, the Department of Defense (DoD) ignited a high-altitude nuclear device 200 miles off the coast of the Hawaiian Islands. An electromagnetic wave which was generated by the high-altitude nuclear device caused extensive damage to the electronics in the Hawaiian Islands. The damage included street lights exploding, many of the powerplants were shut down, and basic communication within the Islands was halted. The DoD realizing that they had a potential military problem turned to the finest electromagnetic theorists in the academic and commercial sector to understand and provide methods to protect their weapon systems from the effects of the electromagnetic wave. This electromagnetic wave was later classified as an electromagnetic pulse (EMP) wave. One of the by-products generated by the investigation was a test method used to test EMI gaskets. This test method became an SAE specification (ARP-1705) [5]. The test method provided accurate test data over the frequency range of 10 kHz to 200 MHz.

The SAE ARP-1705 specification was later updated by EMC engineers to meet the needs of the Electromagnetic Compatibility community. Revision A of the specification provided instructions to (1) test over the frequency range of 10 kHz to 1 GHz with an accuracy of ± 2 dB; (2) test an EMI gasket against the various structural materials and finishes used by industry; (3) the effects force of the gasket against the joint surfaces has on the level of shielding the gasket can provide; and (4) the effects humidity and salt fog environments have on the ability of the gasket to provide adequate shielding. Revisions B and C were added to the specification which provided accurate data to a level of ± 2 dB over the frequency range of 10 kHz to 18 GHz.

SAE ARP-6248 is a method of testing EMI gaskets over the frequency range of 100 MHz to 40 GHz [6]. The test method utilizes a strip line fixture that has a fairly large dynamic capability that can be used to discriminate for the shielding value of the various gaskets on the market. The receiving member has significant standing waves associated with the receiving data. In performing the testing with a network analyzer, it is recommended that the averaging feature of the analyzer be used. The use of the averaging feature will allow for fairly straight data traces. Pre-amplifiers should be used to allow for the fairly large dynamic capability.

Slot aperture and nested reverberation rooms are two methods of testing EMI gaskets that are worthy of a standard. Both test methods are capable of providing accurate shielding effectiveness test data over the frequency range of 1 GHz to 40 GHz. The slot aperture test uses a shielded room modified for performing the shielding effectiveness testing contained in MIL–DTL–83528 [4]. Two plates are attached to the hole in the enclosure where a small horizontal gap in the middle of the hole is obtained. A vertically polarized horn antenna is aimed at the gap where reference test data is taken. An EMI gasket under test is then placed between the two plates and a second measurement is taken. The difference in the two readings represents the shielding effectiveness of the gasket. Test method using nested reverberation rooms possess a hole between the rooms. A reference test is taken with the hole open. A cover plate with a gasket under test is attached to the hole and additional test is taken. The difference in the readings is the shielding effectiveness of the gasket under test. In performing slot aperture testing, a tracking generator is used to obtain a constant trace over the frequency range. In performing the nested reverberation room test, spot frequencies are taken where sufficient number of tests at each frequency will yield an accurate statistical level of shielding.

References

1. IEEE 1302 (2008) Guide for the electromagnetic characterization of conductive gaskets in the frequency range of DC to 18 GHz. IEEE Standards Association
2. MIL-STD-285 (1956-Obsolete) Attenuation measurements for enclosures, electromagnetic shielding, for electronic test purposes, US Department of Defense
3. SAE ARP-1173 Rev. A (2013) Test procedure to measure the RF shielding characteristics of EMI gaskets, SAE International
4. MIL-G-83528 Rev. B (1992-Obsolete) Gasketing material, conductive, shielding gasket, electronic, elastomer EMI/RFI general specification, US Department of Defense
5. SAE ARP-1705 Rev. C (2017) Coaxial test procedure to measure the RF shielding characteristics of EMI gasket materials, SAE International
6. SAE ARP-6248 Rev. A (2004) Corrosive controlled and electrical conductivity in enclosure design, SAE International

Chapter 15
Transfer Impedance Testing of EMI Gaskets

Transfer impedance testing of EMI gaskets is performed per SAE ARP-1705 [1]. The original specification was prepared by EMP engineers. The purpose of the specification was to test EMI gaskets for protecting DoD equipment from the devastating effects of an EMP wave. This specification tested the gaskets over the frequency range of 10 kHz to 200 MHz.

The specification was updated to revision A by EMC engineers to test the gaskets for compliance to the various EMC requirements. The frequency range is from 10 kHz to 1.0 GHz with an accuracy of ± 2 dB [2–3]. The revision included the following test requirements:

1. A quality control test using gold plated test plates.
2. A test where the test plates were constructed from the material and finish to be used in actual application of the gasket under test.
3. An environmental set of tests using test plates constructed from the various materials and finishes to be used in actual application. Tables 15.1 and 15.2 illustrate the environmental testing to be performed.

Revision C was added to permit testing over the frequency range of 10 kHz through 18 GHz. The testing to revision C has proven to be accurate to a level of ± 2 dB. The testing using the test fixture of 1705C tests for (1) quality control test and (2) testing using the test plates constructed from the materials and finishes used by industry.

The results of the transfer impedance testing are in ohm-meters, where transfer impedance is defined as:

$Z_T = V/J_S$

Z_T = Transfer impedance of seam (Ω-meters)

V = Transfer Voltage (voltage across seam)

J_S = Density of current which crosses the seam (Surface Current Density in A/m)

© Springer Nature Switzerland AG 2020
G. M. Kunkel, *Shielding of Electromagnetic Waves*,
https://doi.org/10.1007/978-3-030-19238-9_15

Table 15.1 Transportation/storage environmental test requirements

Transportation/storage environments	
Environmental stress Condition	Test method/procedure (MIL-STD-810G)
High temperature (Dry/humid)	Method 501
Low temperature (Rain/hail/freezing)	Method 502
Thermal shock	Method 503
Solar radiation	Method 505
Fungus growth	Method 508
Rain	Method 506
Humidity	Method 507
Salt fog	Method 509

Table 15.2 Mission/sortie environmental test requirements

Mission/sortie environments	
Environmental stress Condition	Test method/procedure (MIL-STD-810G)
High temperature (Dry/humid)	Method 501
Humidity	Method 507
Salt fog	Method 509
Explosive atmosphere	Method 511
Rain	Method 506
Emersion	Method 512

The transfer impedance test data can be converted to a quantity defined as shielding quality. The value of shielding quality is obtained by dividing the impedance of the incident wave by the transfer impedance of the gasket under test (the impedance of the wave is assumed to be 377 ohms unless otherwise stated). This figure of merit is within a few dB from the predicted shielding effectiveness of the gasket.

Figure 15.1 illustrates the basic components of the 1705A test fixture.

Calibrated current from a tracking generator is brought into the transfer impedance test fixture through the 50 ohm resistor assembly. The current is deposited on the contact plate and flows out to the gasket under test. It crosses the gasket under test to the base plate and returns to the 50 ohm resistor assembly. The voltage across the gasket under test is measured with a spectrum (or Network) analyzer attached to the receiver pin assembly.

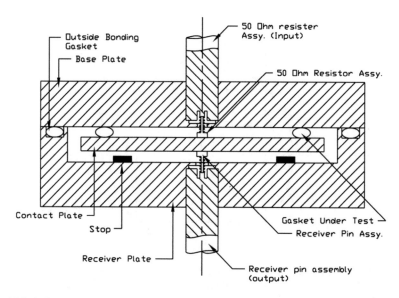

Fig. 15.1 1705A test fixture

The transfer impedance of the gasket is obtained by dividing the voltage across the gasket by the current which is delivered to the 50 ohm resistor assembly and multiplying the results by the circumference of the gasket in meters. The calibrated current is obtained by dividing the input voltage by 50 ohms.

Figure 15.2 shows a base plate illustrating the maximum size of a gasket that can be tested using the 1705A test fixture.

As is illustrated the maximum size of a square gasket is 3.4 in. and the maximum diameter is 4 in. The maximum height of any gasket (assuming a 25% compression) is three-eighths of an inch.

In performing a quality control test, the contact and base plates are constructed from 6061–T6 aluminum, with the contact surfaces plated with gold. In performing all other tests, the contact and base plates are constructed and plated from the material and finish of interest.

The base and contact plates are held together using ¼–20 nylon cap screws. The spacing between the plates is controlled with nonconductive washers. The nominal compression of the gaskets under test is 25%. It is often desirable to test for compression of less than 25%. During this testing, the percent compressed is controlled by nonconductive washers.

When performing the environmental testing, a nylon plug is screwed into the top of the base plate to prevent moisture or salt-fog environments from entering the

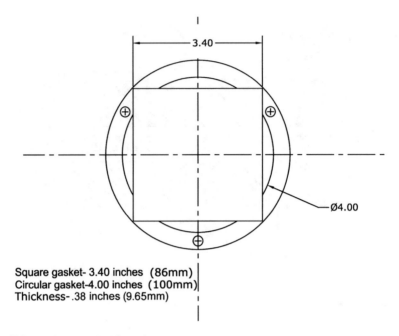

Fig. 15.2 Maximum gasket dimensions

plates through the hole. It is recommended that a rubber o-ring be used in conjunction with the nylon plug to ensure moisture and salt-fog environments do not enter the test plates through the hole. Figure 15.3 illustrates the 1705C test fixture.

The transfer impedance testing using the 1705C test fixture is identical to that of the 1705A test fixture. The ability to test to 18 GHz is achieved by reducing the size of the gasket under test to a maximum diameter of 0.75 in. The ability to maintain a ±2 dB accuracy in the test data is achieved through the use of the adapter sleeves associated with both the input and output adapters. The sleeves are used to maintain a 50 ohm characteristic impedance of the 50 ohm resistor assembly to the gasket under test. A sleeve is also used to maintain a 50 ohm characteristic impedance from the gasket to the receiver assembly.

The size of the gasket under test has a diameter of 0.75 in. by a maximum height of 11/64s of an inch.

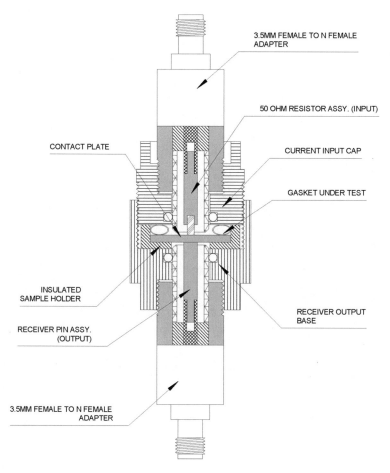

Fig. 15.3 1705C test fixture

References

1. SAE ARP-1705 Rev. C (2017) Coaxial test procedure to measure the RF shielding character-istics of EMI gasket materials, SAE International
2. Freyer GJ (1992) Comparison of gasket transfer impedance and shielding effectiveness mea-surements, part I, IEEE, EMC International Symposium, Anaheim, CA
3. Hatfield MO (1992) Comparison of gasket transfer impedance and shielding effectiveness measurements, part II, IEEE, EMC International Symposium, Anaheim, CA

Chapter 16
Shielding Effectiveness Testing of Shielding Components: Paint, Glass, and Air Vent Materials

Test Fixture

The shielding effectiveness test fixture consists of four different assemblies. These assemblies are (1) 50 ohm transmitting assembly as illustrated in Fig. 16.1, (2) receiving assembly as illustrated in Fig. 16.1, (3) test plate assembly, and (4) spacer assembly.

The 50 ohm transmitting assembly consists of (1) tin plated 50 ohm input base, (2) female to female type N adapter, (3) four 200 ohm film resistors in parallel – equaling 50 ohms, (4) 3.0 in. diameter brass transmitting plate, (5) positioning spacer, (6) 0.058 in. diameter pin, and (7) groove mounted 9/64 diameter EMI gasket.

The receiving assembly consists of (1) tin plated receiving output base, (2) female to female type N adapter, (3) 3.0 in. diameter brass receiving plate, (4) phenolic alignment spacer, (5) 0.058 in. diameter pin, and (6) groove mounted 9/64 in. diameter Electromagnetic Interference (EMI) gasket.

The test plate assembly consists of (1) tin plated test plate with 3.0 in. square hole and applicable threated screw holes, (2) phenolic force plate, and (3) groove mounted Electromagnetic Interference (EMI) gasket.

An EM bonding spacer assembly is to be used when testing 1.0 in. thick honeycomb panels. The assembly consists of (1) 0.5 in. thick bonding spacer and (2) groove mounted 9/64 in. diameter EMI bonding gasket.

Framing Fixture

A framing fixture is used to hold the test fixture together and provide a vertical force to EM bond the assembly together.

© Springer Nature Switzerland AG 2020
G. M. Kunkel, *Shielding of Electromagnetic Waves*,
https://doi.org/10.1007/978-3-030-19238-9_16

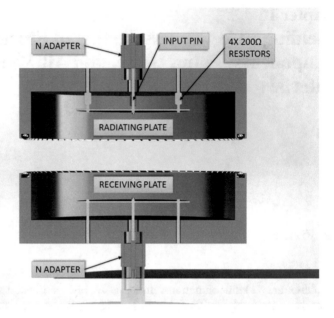

Fig. 16.1 Test fixture assembly

Theory of Operation

1. Power from a tracking generator over the frequency range of 100 kHz to 1.0 GHz is supplied to the 50 ohm transmitting assembly. The 50 ohm termination reduces (to the greatest extent possible) standing waves. This results in smooth data traces. The use of amplifies and pre-amps permits a dynamic range at 1.0 GHz of 140 dB.
2. The current that is delivered to the transmitting plate flows out to the resistors and back to the source of power. This current flow creates a voltage differential across the resistors, where lines of flux radiate out from the transmitting plate.
3. A reference level test is taken with the hole in the test plate open.
4. The sample under test is then attached to the test plate and an additional test is observed and recorded.
5. The difference in the recorded data is the shielding effectiveness of the test sample.
6. The recorded data is obtained with the use of a spectrum (or network) analyzer attached to the receiver assembly.

Chapter 17
Transfer Impedance Testing of Shielded Cables, Back Shells, and Connectors

Overview

Cable shielding: The level of shielding offered by a shielded cable is a function of (1) the methods used in the manufacture of the shield and (2) the method used to reference the shield to ground. The actual shielding offered by the shield on a cable can vary between close to zero (when the length of the cable is equal to 1/2 the wavelength of the signal being carried in the cable) to as much shielding as required at any frequency or range of frequencies. Referencing of the ends of the shield to ground is critical. Back shells have been developed to provide a positive ground for a shielded cable [1].

The preferred ground reference path for the shield on a cable is through a back shell to the interface connector and then to ground. The interface connector consists of two (2) parts. This is a connector which is tied to the cable and a receptacle which is attached to the chassis of a system or subsystem. The connector pair which is used to reference the shield of the cable to ground is so designed as to provide an electromagnetic bond between the connector and the receptacle. If an EMI gasket is used to reference the receptacle to ground the specific type of gasket is to be documented and is part or the assembly.

Test variables: In performing the transfer impedance testing of the shield on a cable, the back-shell assembly, and the shielded cable assembly, two (2) shielded enclosures referenced together on a ground plane are to be used as illustrated in Fig. 17.1. In EM bonding a test sample to one of the enclosures, the EM bond must be well designed to provide a positive ground.

In the performance of the testing, the wire or wires contained in the test sample are to be tied to the spectrum or network analyzer voltage probe at one end and a 50 ohm termination at the other end. In obtaining the induced voltage, all of the wires in a cable should be tied together. However, a close estimate of the induced voltage can be obtained using a single wire and multiplying the induced voltage by the

© Springer Nature Switzerland AG 2020
G. M. Kunkel, *Shielding of Electromagnetic Waves*,
https://doi.org/10.1007/978-3-030-19238-9_17

Fig. 17.1 Test configuration

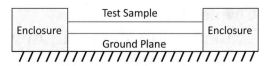

square root of the number of wires in the cable. The current induced onto the test sample is to be injected using a current probe placed on the test sample as close to the enclosure containing the voltage probe as possible. A secondary current probe is to be placed next to the injection probe as possible to measure the current injected onto the test sample.

The transfer impedance of the test sample under test is the voltage induced onto the wire or wires divide by the current induced onto the test sample. The inverse of the transfer impedance test data (when both ends of the wire or wires are terminated into 50 ohms) is a close estimate of the shielding effectiveness of the test sample.

Test Outline

Shielded cable: Testing to be performed is as follows:

1. Reference the ends of the cable under test to both of the shielded enclosures. This reference is to be a solid 360° bond.
2. Attached the spectrum or network analyzer voltage probe and 50 ohm termination to the wire or wires of the cable inside the enclosures.
3. Measure and record the transfer impedance of the cable.

Back shell and connector assembly: Testing to be performed is as follows:

1. A tin-plated aluminum tube is to be used to simulate a shielded cable. This tube is to be a minimum of 2 ft. in length.
2. One end of the tube is to be attached to the back shell under test. If the shield of the cable is to be used to reference the cable to the back shell, the shield of the cable is to be soldered onto the tube.
3. The other end of the tube is to be referenced to the other shielded enclosure. This reference is to be a solid 360° bond.
4. A wire is to be inserted in the center of the tube with one end attached to the spectrum or network analyzer voltage probe and the other end to a 50 ohm termination.
5. The wire is to be held in the center of the tube using Styrofoam material.
6. The back shell is to be attached to the appropriate connector with the receptacle attached to the shielded enclosure in which the voltage is to be measured.
7. Observe and record the transfer impedance of the back shell and connector assembly.

Quality Control Test of Shielded Cable Assembly

Testing is to be performed as follows:

1. The shielded cable assembly consists of the shielded cable, the back shell, and connector attached to both ends of the cable.
2. The receptacle associated with the connector pair is attached to both of the shielded enclosures as to be used in actual application.
3. The spectrum or network analyzer voltage probe and 50 ohm termination are to be attached to the wire or wires of the cable assembly.
4. Observe and record the transfer impedance of the cable assembly.

Reference

1. Szentkuti BT (1992) Shielding quality of cables and connectors: some basics for better understanding of test methods. IEEE, EMC International Symposium, Anaheim, CA

Appendix A: Critique of MIL-DTL-83528G

Abstract MIL-DTL-83528 is a general specification to be used by all departments and agencies of the Department of Defense (DoD) for purchasing the silver elastomeric electromagnetic interference (EMI) gasket for providing environmental and shielding protection for DoD weapon systems. The critique illustrates that the gasket material can have a short storage and useful life. The products are incompatible with joint surfaces when exposed to a moisture and salt fog environment as stipulated in an SAE standard[1] and paragraph 6.1.2 of MIL-DTL-83528G. This lack of compatibility results in a significant loss of shielding. There are two general paragraphs describing the elastomeric products that need revisions detailing that the materials are incompatible with moisture and salt fog environments. The shielding effectiveness test contained in the specification is designed to produce highly inflated test results. The critique illustrates the results of two tests. One test is of a newspaper that possesses a shielding effectiveness of 93 dB at 2 GHz and the test results of a nonconductive phenolic gasket that shows 68 dB of shielding at numerous frequencies between the frequency range of 20 MHz and 10 GHz.

MIL-DTL-83528 is a general specification to be used to purchase silver elastomeric EMI/RFI gaskets to provide environmental and shielding protection for DoD weapon systems [1].

There are two (2) significant problems associated with the silver elastomeric gasket for use in DoD weapon systems. The problems are:

1. The elastomeric gasket can possess a short shelf life. Paragraph 6.6 of the specification provides detailed information on how to store the gaskets (to protect the gasket material from the storage area environment). If properly stored, the gaskets can have a shelf life of 15 years. However, if the elastomeric gasket is not stored as stipulated, the shelf life can be as little as 4–6 months.
2. The silver elastomeric gasket when designed into systems that are to operate in humid or salt fog environments, the shielding will be highly compromised. Paragraph 6.1.2 of the specification states that the silver elastomeric gasket is incompatible with flange surfaces in the presence of a salt fog environment. The

---

[1] SAE International, initially established as the Society of Automotive Engineers, is a U.S.-based, globally active professional association and standards developing organization for engineering professionals in various industries. www.sae.org.

© Springer Nature Switzerland AG 2020
G. M. Kunkel, *Shielding of Electromagnetic Waves*,
https://doi.org/10.1007/978-3-030-19238-9

matrix contained in SAE ARP-1481 illustrates that the silver elastomeric gaskets are incompatible with most flange surfaces in the presence of a humid environment.

During the 1950s, the silver elastomeric gasket was introduced to the marketplace. The product was being marketed as a lightweight, environmental gasket providing "exceptional" shielding protection. This exceptional shielding protection (as advertised) was made possible by introducing a shielding effectiveness test method that produced excessively high levels of shielding. The qualities of the gasket as advertised appeared desirable to the Air Force, where they designed the gasket into a large number of their weapon systems. When the patent protection expired, the company that manufactured and introduced the silver elastomeric gasket convinced the Air Force that an Air Force specification was required to ensure that other companies manufacturing a like product would produce products as good as or better than those designed into their weapon systems. The result was an Air Force specification (MIL-G-83528). This specification was prepared by the company that manufactured the silver elastomeric gasket, where the resultant specification appeared highly complementary in the use of the gasket. The specification later became a DoD specification (MIL-DTL-83528). Numerous revisions to this specification have been prepared where many of the paragraphs appear to have enhanced the properties of the silver elastomeric products. The original document and the apparent enhancements should be revised to produce a document that truly represents the elastomeric products contained in the specification.

The paragraphs and sentences of MIL-DTL-83528 that need revision, the justification for the need, and the recommended revisions are as follows.

Paragraph 4.5.12a

"*A relative measurement of the shielding effectiveness of the material shall be made in accordance with a documented method acceptable to the qualifying activity (see 6.1.4).* Shielding effectiveness shall be defined as the ability of a gasket material to electrically bond a test cover panel to an enclosure flange such that radiated RF through a 24-by-24-in (61-by-61 cm) opening is attenuated by the factors specified. The test configuration of figure 4 in MIL-DTL-83528 will provide more than 120 dB of dynamic range (E-field) through the 24-by-24-in (61-by-61 cm) opening for frequencies above 20 MHz. Swept frequency techniques are encouraged, but as a minimum, data shall be recorded at the 1, 2, 4, 6, and 8 times frequencies of each decade in the 20 MHz through 10 GHz range. The position of antennas, equipment, or other metal-containing objects in the shielded room should not be moved between open-aperture and closed-aperture measurements. An optional shielding effectiveness test can be conducted with the transmitting antenna inside the enclosure and the receiving antenna outside the enclosure and sufficient dynamic range can be achieved at

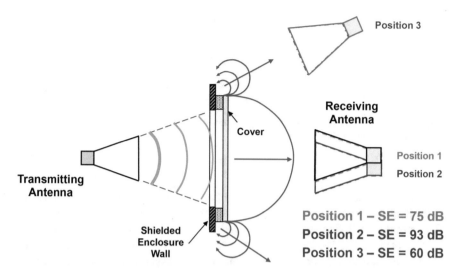

Fig. A.1 MIL-DTL-83528C shielding effectiveness testing of newspaper at 2.0 GHz

all frequencies. NOTE: The enclosure must be large enough that no part of the transmitting antenna is within one meter of any enclosure surface."

Paragraph 4.5.12a Comments

The first sentence of the paragraph states: "A relative measurement of the shielding effectiveness of the materials shall be made in accordance with a documented method acceptable to the qualifying activity (see 6.1.4)." Paragraph 6.1.4 states: "IEEE-STD-299 has been considered an acceptable method for determining shielding effectiveness as modified and supplemented by figure 4."

The test method is a modified MIL-STD-285 (IEEE-299) test method [2]. These modifications have proven to produce inflated test data of as much as 60 dB and are [3]:

1. Aiming the transmitting antenna at the center of a large 26″ × 26″ aluminum plate instead of at the gasket seam of the gasket under test.
2. The receiving antenna is aimed at the center of the large 26″ × 26″ aluminum plate instead of measuring the maximum field strength radiating from the gasketed seam.

Figure A.1 illustrates the results of a shielding effectiveness test[2] performed on a newspaper at 2 GHz using the shielding effectiveness test method of figure 4.

As is illustrated in Fig. A.1, the recorded shielding effectiveness varied between 60 dB and 93 dB. The 33 dB difference was obtained due to the different location of

---

[2] Note: the test was performed by Al Broadus, President of DNB Engineering in Fullerton California and reported during the 1983 IEEE, EMC International Symposium.

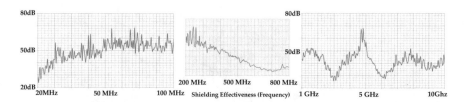

Fig. A.2 Shielding effectiveness test data performed per MIL-DTL-83528 Rev. G on a 1/8 in. thick (non-conductive) phenolic gasket

the receiving antenna (the difference in taking a measurement with the receiving antenna aimed at the center of the aluminum plate versus measuring the maximum field strength radiating from the gasketed joint). The predicted difference in the recorded shielding effectiveness due to aiming the transmitting antenna at the center of the aluminum plate versus at the seam of the newspaper is greater than 30 dB, that is,

The impedance associated with the capacitive reactance of a newspaper 0.001 in. thick results in a predicted shielding effectiveness of 24.6 dB.

With regard to this newspaper shielding effectiveness test, the inflated data due to the revisions resulted in inflated test data of more than 60 dB [4].

Figure A.2 illustrates the test results of measuring and recording the shielding effectiveness of a nonconductive 1/8 in. thick phenolic gasket[3] over the frequency range of 20 MHz to 10 GHz using the test method of figure 4 [3].

As is illustrated the shielding effectiveness varied between 34 dB and 68 dB. The predicted shielding effectiveness over that frequency range is 4 dB. Thus, the inflated data varied between 30 dB and 64 dB.

Based upon the above shielding effectiveness test data, it is obvious that the qualifying activity is not technically qualified to make the statement contained in paragraph 6.1.4. As such the first sentence contained in paragraph 4.5.12a and the contents of paragraph 6.1.4 should be removed from the specification.

Paragraph 4.5.12b.

"It may not be inferred that the same level of shielding effectiveness provided by a gasket material tested in the enclosure of figure 4 would be provided in an actual equipment flange, since many mechanical factors of the flange design (tolerances, stiffness, fastener location, and size) will affect shielding effectiveness. This procedure provides data applicable only to the test enclosure and cover panel design of

---

[3] Note: the test was performed at DNB Engineering in Fullerton California using their anechoic chamber modified to test for the shielding effectiveness as contained in MIL-DTL-83528 [4].

Figure 4, but which is useful for making comparisons between different gasket materials."

Paragraph 4.5.12b Comments.

The first sentence of the paragraph states that the test data may not be realized in actual application, for totally irrelevant reasons. The test method was originally designed to provide highly inflated test data for the silver elastomeric gasket for marketing purposes. This test method was lobbied into MIL-G-83528. This test standard was revised and replaced with MIL-DTL-83528, where the test method contained in the new standard is in all practical purposes identical to the original test procedure.

The first sentence should be changed to read:

"The data obtained using the enclosure of Figure 4 will not be realized in actual practice, where the difference can exceed 60 dB."

Paragraph 6.1.1

"General. *Gaskets covered by this specification are designed to provide EMI/RFI shielding, EMP survivability, and environmental protection for electronic enclosures, connectors, and waveguides.* Their principal areas of application are aircraft, missiles, spacecraft, and ground support equipment. This does not preclude the use of these gaskets in other military applications."

Paragraph 6.1.1 Comments

The paragraph states that the gaskets are designed to provide EMI/RFI shielding for various military applications. This statement implies that the shielding will be obtained under all conditions.

Paragraph 6.1.2 of the specifications states that the elastomeric gaskets are incompatible with certain flange surfaces in the presence of a salt fog environment. The matrix of SAE ARP-1481 stipulates that the gaskets are incompatible with most joint surfaces in the presence of a humidity environment. In fact, the silver elastomeric gasket is less compatible with the various materials and finishes used by industry than any other gasket on the market in humid and salt fog environments due to its silver content (see SAE ARP-1481) [5].

A note should be added to the paragraph that states: "The EMI/RFI protection cannot be achieved when in contact with most of the structural materials and finishes used by industry in the presence of moisture and salt fog environments."

Table A.1 Humidity compatibility of silver elastomeric gaskets

Finish	Aluminum series 1000, 2000, 3000, 5000, 6000 & 7000
None	Requires sealing regardless of exposure
AMS-C-5541 Class 1A	Requires sealing if exposed to humidity (or salt fog) environments
AMS-C-5541 Class 3	Requires sealing if exposed to moisture (or salt fog) environment
Nickel	Compatible in environment of controlled temperature and humidity only

Paragraph 6.1.2

"**Salt spray environments**. *All EMI gasket materials (metal and elastomer) to varying degrees are incompatible with certain flange surfaces.* Design of the joint, therefore, plays a central role in determining the electrical stability and corrosion resistance of the joint. Design variables include: Flange material and finish, gasket filter and form (sheet, O-ring in a groove), use of parallel nonconductive environmental gaskets, mechanical design, and use of insulating coatings. Choice of the design options should depend on: Environment of the application, levels of shielding effectiveness required versus frequency, and expected life of the equipment. When designing for salt spray environments, all of the preceding factors must be considered."

Paragraph 6.1.2 Comments

The first sentence states: "All EMI gasket materials (metal and elastomeric) are incompatible with certain flange surfaces." The sentence is correct. However, the "metal" EMI gasket materials that are incompatible with certain flange surfaces are protected by an environmental seal. The elastomeric material (covered by the standard) functions as an EMI and Environmental Seal (see paragraph 6.1.1). As such, the metal EMI materials are not exposed to a salt spray environment. The silver elastomeric materials are exposed to the salt spray environment and are not compatible with certain flange surfaces.

The title of the paragraph should be changed to "Moisture and Salt Fog[4] Environments." The contents of the paragraph should state: "The silver elastomeric materials covered by this specification are to be sealed to protect the gasketed flanges from moisture and salt fog environments."

Moisture is included per the requirements of the Society of Automotive Engineers SAE ARP-1481[5] entitled "Corrosive Controlled & Electrical Conductivity in Enclosure Design." Included in the ARP is a matrix containing six EMI gasket

---

[4] Salt Spray is a test, Salt Fog is an environment.

[5] Note: SAE ARP-1481 was prepared by a committee chaired by Earl Grossart. Mr. Grossart was an environmental engineer employed by the Boeing Aerospace Group and was responsible for Corrosion Compatibility of the systems designed and built by the group [5].

materials (including silver elastomeric), listing the compatibility of each of the gaskets as a function of various structural materials and finishes. Table A.1 lists the degree of compatibility of the silver elastomeric gasket against aluminum and subsequent finish.

Paragraph 6.6

"**Storage**. Material should be stored in sealed polyethylene when possible; otherwise, it should be stored in such a way that it is not exposed to sulfur. Sulfur-cured materials or materials containing sulfur based plasticizers (such as most neoprenes) should not be stored in close proximity to materials covered by this specification. When stored between 50°F and 90°F, in cabinets, bins or any other storage container which prevents excessive exposure to light, and in the absence of sulfur, the shelf life should exceed 15 years."

Paragraph 6.6 Comments

The paragraph lists methods of storing for the materials covered by the specification. The paragraph should be relabeled "Packaging and Storage." The contents should address a method of packaging that will protect the material from transportation and storage environment. The recommended contents are: "The materials covered by this specification shall be packaged in polyethylene (or equivalent) sealed packaging materials with zip-lock capability." A warning sign should be applied to the packaging stating: "The material contained herein shall be stored in this packaging to protect the contents from the storage environment to ensure storage life."

Paragraph 6.7.1

"**Conductive elastomer EMI gaskets**. *Conductive elastomer gaskets are highly electrically conductive, mechanically resilient and conformable vulcanized gaskets which provide low interface resistance between mating electronic enclosure flanges or covers while simultaneously providing moisture, pressure, or environmental sealing.*"

Paragraph 6.7.1 Comments

A sentence at the end of the paragraph should be added to state: "The edge of the gasket material exposed to the environment is to be sealed to prevent moisture and salt fog environments from creating galvanic corrosion resulting in a significant loss of conductivity."

References

1. MIL-DTL-83528 Rev. G (2017) Gasketing material, conductive, shielding gasket, electronic, elastomer EMI/RFI general specification, US Department of Defense
2. MIL-STD-285 (1956-Obsolete) Attenuation measurements for enclosures, electromagnetic shielding, for electronic test purposes, US Department of Defense
3. Kunkel G (2018) Testing of EMI gaskets. Evaluation Engineering Magazine
4. Broaddus Al, Kunkel G (1983) Shielding effectiveness test presentation, IEEE, EMC International Symposium, Arlington, VA
5. SAE ARP-1481 Rev. A (2004) Corrosive controlled and electrical conductivity in enclosure design, SAE International

Appendix B: Maxwell's Field Equations

$$\mathrm{rot}\,\mathbf{H} = i + \frac{\partial \mathbf{D}}{\partial t}$$

$$\mathrm{rot}\,\mathbf{E} = -\frac{\partial \mathbf{B}}{\partial t}$$

$$div\,\mathbf{B} = 0$$

$$div\,\mathbf{H} = \rho_m$$

$$div\,\mathbf{D} = \rho$$

$$\begin{cases} \mathbf{B} = \mu_0(\mathbf{H} + \mathbf{M}) \\ \mathbf{B} = \mu\mu_0\mathbf{H} \end{cases}$$

$$\begin{cases} \mathbf{E} = \dfrac{1}{\varepsilon_0}(\mathbf{D} - \mathbf{P}) \\ \mathbf{E} = \dfrac{1}{\varepsilon\varepsilon_0}\mathbf{D} \end{cases}$$

$$\mathbf{i} = \sigma\,\mathbf{E}$$

$$div\,\mathbf{i} + \frac{\partial \rho}{\partial t} = 0$$

$$\frac{dW}{dv} = \tfrac{1}{2}\mathbf{E}\cdot\mathbf{D} + \tfrac{1}{2}\mathbf{B}\cdot\mathbf{H}$$

E: electric field strength (V/m)
B: magnetic flux density (Wb/m^2)
D: electric flux density (C/m^2)

© Springer Nature Switzerland AG 2020
G. M. Kunkel, *Shielding of Electromagnetic Waves*,
https://doi.org/10.1007/978-3-030-19238-9

H: magnetic field strength (A/m)
ρ = electric charge density
ρ_m = magnetic charge density
i = electric current density
ε_0 = permittivity/dielectric constant
J = current density
μ_0 = permeability
c = speed of light
M = magnetization
P = electric polarization

Reference

1. Hallen E (1962) Electromagnetic theory. Wiley, New York

Appendix C: Shielding Theory Equations

$$SE = R + A + B\,(dB)$$

where:

$$R = 20\log\frac{(K+1)^2}{4|K|} \ \text{Reflection loss}\,(dB)$$

$$A = 8.686\alpha d \ \text{Absorption loss}\,(dB)$$

$$B = 20\log\left|1 - \left(\frac{K-1}{K+1}\right)^2 e^{-2\alpha d}\right| \ \text{Reflection correction factor}\,(dB)$$

$$K = \frac{Z_{\text{wave}}}{Z_{\text{barrier}}}, \ Z_{\text{barrier}}\left(\frac{j\omega\mu}{\sigma}\right)^{1/2} = \frac{1+j}{\sigma\delta}$$

$$
\begin{aligned}
Z_{\text{wave}} &\approx -j377\,\lambda/2\pi r,\,(r < \lambda/2\pi) \ \text{High impedance source}\\
&\approx j377(2\pi r/\lambda),\,(r < \lambda/2\pi) \ \text{Low impedance source}\\
&\approx 377,\,(r \geq \lambda/2\pi) \ \text{All sources}
\end{aligned}
$$

$$\alpha = \left(\frac{\mu\sigma\omega}{2}\right)^{1/2} = 1/\delta$$

d = Thickness of Barrier (m)
  r = Distance from Source to Barrier (m)
  $\omega = 2\pi f$
  μ = (Absolute) Permeability of Barrier (H/m)

© Springer Nature Switzerland AG 2020
G. M. Kunkel, *Shielding of Electromagnetic Waves*,
https://doi.org/10.1007/978-3-030-19238-9

$$\mu = 4\pi \times 10^{-7}$$

σ = (Absolute) Conductivity of Barrier (mhos/m)

$$\sigma_{copper} = 5.82 \times 10^{7}$$

$$\sigma_{aluminum} = 3.55 \times 10^{7}$$

$$\lambda = c \,/\, f = 3 \times 10^{8} \,/\, f\,(\mathrm{m})$$

Reference

1. Frederick Research Corp (1962) Handbook of radio frequency interference, vol 3 (Methods of Electromagnetic Interference Suppression). Frederick Research Corp., Wheaton, MD

Index

© Springer Nature Switzerland AG 2020
G. M. Kunkel, *Shielding of Electromagnetic Waves*,
https://doi.org/10.1007/978-3-030-19238-9

Printed in the United States
By Bookmasters